Neural Powerhouse: Unveiled

Vani

Contents

CONTENTS

CONTENTS

CONTENTS

Chapter 1

Introduction

1.1 Brain Disease

The human brain is a very complex organ that controls bodily activities and systems. It uses about 20% of the body's energy and contains a total of 100 billion neurons supported by 1,000 billion glial cells. The cerebral cortex, which is the outer layer of the brain, contains about one-sixth to one-third of the total neurons in the brain. The brain has four main lobes; these are temporal, occipital, parietal and frontal lobe. It consists of many regions, which include ganglia, cerebellum, Broca's area, corpus callosum, medulla oblongata, amygdala, hypothalamus and thalamus. Various brain diseases might affect different regions of the brain, and depending on the function mapped to that region, bodily activities are affected, and certain symptoms arise (Herculano-Houzel, 2009).

Brain diseases come in many forms; they include traumas, seizures, strokes, tumors and brain cancers, neurodegenerative conditions and psychiatric disorders. Neurodegenerative diseases affect the brain cells and cause neuron death and damage. Examples of neurodegenerative disorders include amyotrophic lateral sclerosis, dementia, Alzheimer's, Parkinson's and Huntington's diseases. Brain tumors can be benign or malignant, and are caused by the abnormal growth of cells in the brain. Benign tumors are mostly slow growing and they can originate in different parts of the brain. The most common type of benign tumors are meningiomas, and they start growing from the lining of the brain. Malignant brain tumors, on the other hand, are aggressive, and they invade nearby tissues (e.g. glioblastoma). Other types of brain diseases include viral or bacteria infections, which can cause inflammation around, or within, the brain (e.g. meningitis and encephalitis). Traumas are injuries that disturb the brain function. One type of trauma is concussion, which affects the cognition ability and causes confusion and unconsciousness. Intracerebral haemorrhage is a different type of trauma which causes bleeding in the brain. Autoimmune conditions, such as multiple sclerosis, are other types of brain disorders that cause the immune system to attack the nerve cells (McDonald et al., 2001; Goetz, 2007).

Here, we focus on studying two main aspects of brain diseases, that is, genetics and neuroimaging characteristics, as well as studying their complex interplay and relationship with other environmental and clinical factors. Specifically, we intend to extract neuroimaging phenotypes of brain structure and study their causal relationship with some *multi-omic* data. Multi-omic data refer to the information extracted from either the genome, proteome, transcriptome or metabolome, we focus here on gene expressions and single nucleotide polymorphisms. In this book, we decided to narrow our focus to one focal lesion disease, namely glioblastoma multiforme, and one disease affecting more widespread regions across the brain, namely Alzheimer's disease. This way we will be able to study different structures of the brain, identify potential genetic risk factors which are responsible for changes in brain structure, and get easy access to imaging genetics datasets. We studied the two diseases and tested different hypotheses to understand the association be-tween neuroimaging characteristics and multi-omic factors. In doing so, we identified the need for a unified model to study the complex interplay between genetic, environmental and clinical, neuroimaging and phenotype features. We introduce a novel model which can test complex hypotheses in the field of imaging genetics, study the effect of interaction terms on the final phenotype, and accommodate heterogeneous data types including phenotype measurements, brain connectivity, environmental and multi-omic factors.

1.2   Alzheimer's Disease

1.2.1 Clinical Characteristics. Alzheimer's Disease (AD) is a central nervous system degenerative disorder that occurs in the brain, and worsens over time. It is suspected to develop 20 years before the disease symptoms appear (Braak et al., 2011; Reiman et al., 2012; Gordon et al., 2018). Specifically, AD is thought to start by slight changes in the patient's brain; however, those changes are not noticeable at first. Eventually, after years, the nerve cells in the brain (or; neurons) start to die in different regions of the brain. Consequently, the disease symptoms start to arise. As a result of the amount of change that occurs in the brain, caused by the death and destruction of the neurons, the symptoms worsen and differ. This results in cognitive decline in many functions, including memory, learning and thinking (Gaugler et al., 2019).

Symptoms of Alzheimer's Disease

Symptoms of AD start to occur as the brain cells, or neurons, involved start to die or damage. As the disease progresses, neurons in different regions in the brain continue to die, and symptoms worsen. The early symptoms of AD mainly affect the cognitive function of patients. These include memory loss of recent events, problem-solving and task management difficulties, confusion of dates and times, poor judgement, misplacing things, changes in personality, and being socially inactive. Eventually, the symptoms start to affect the patients' bodily functions, such as walking and swallowing. Ultimately, AD leads to death (Gaugler et al., 2019).

Stages of the Disease

There are three stages of AD, namely; preclinical AD (Sperling et al., 2011), mild cognitive impairment (MCI) due to AD (Albert et al., 2011) and dementia due to AD (McKhann et al., 2011). Although research is still ongoing to understand and diagnose the preclinical stage, the symptoms of AD are not observable at this stage. However, perceptible changes, or biomarkers, in the brain, blood and cerebrospinal fluid start to become noticeable. In the MCI and dementia due to AD stages, symptoms are present and observable, these include subtle cognitive decline, e.g. thinking disabilities, as well as further changes in brain, such as elevated levels of beta-amyloid (or Amyloid beta; $A\beta$)[1] protein. The symptoms do not interfere much in everyday activities at the MCI stage, and patients can function independently, but they worsen at the stage of dementia, when the decline in daily activities, cognitive and behavioural disabilities become observable. These worsen as the disease progresses, and depend on the degree of damage in the neurons around the brain. Even though 60% to 80% of dementia cases are due to Alzheimer's, not all dementias are caused by Alzheimer's, nor will all individuals diagnosed with MCI develop AD (Wilson et al., 2012). Nevertheless, people who are diagnosed with MCI and have memory problems are more likely to develop AD and dementia (Kantarci et al., 2009).

Statistics and Facts

The progression of AD is somehow subtle and slow. The average postdiagnosis survival duration of an AD dementia patient who is aged 65, or older, is about 4 to 8 years - with no significant differences between males and females or between ethnicity. There are, however, some patients who live up to 20 years after diagnosis. The average mortality rate was shown to be 10.7 per 100 person-years in AD, that is, in every 100 individuals in the population 10.7 are expected to

---

[1]See Section 1.2.3 and Section 1.2.4 for more details about $A\beta$.

die every year, and is higher among the elderly (Brookmeyer et al., 2002; Helzner et al., 2008). Among all causes of death in the category of the population of age more than 80 years, 61% are expected to be in AD patients, while they form 30% of deaths in the general population. Between the age of 70 to 80 years, AD patients spend 40% of the time in the most severe stage of dementia, while they spend 30% of the time in the mild and moderate stages (Arrighi et al., 2010). In the United States, AD is the fifth-leading cause of death for those older than 65 years. Overall, AD is the sixth highest cause of death after heart disease, cancer, unintentional injuries, chronic lower respiratory disease and stroke, ranked according to the number of deaths (Murphy et al., 2018). The increase in AD death rate since 2000 is 31% for those aged from 65 to 74 years, 57% for 75 to 84 years old, and 86% for patients older than 85 years (for Disease Control et al., 2018).

According to the 2019 Alzheimer's Disease Facts and Figures, reported by Gaugler et al. (2019), the deaths in AD might be more than what has been reported by official sources. Patients live with the disease as it progresses, or they die because of it. Therefore, AD increases the status of poor health and morbidity.

Diagnosis

Several factors are being used as biomarkers to diagnose AD at its different stages. Depending on the biomarker of interest, the method of diagnosis is determined. In the preclinical stage of AD and the more advanced stages, a number of biomarkers are present and detectable. The most important of these biomarkers is the amount of $A\beta$ and tau proteins in the brain. The former can be measured through positron emission tomography (PET) amyloid imaging[2]. PET imaging uses functional nuclear techniques to monitor the metabolic process in the brain or other organs in the body. The level of some proteins (e.g. tau and $A\beta$) in fluids, such as the cerebrospinal fluid (CSF) can be detected through the CSF assays (Sperling et al., 2011).

Another biomarker is the level of glucose metabolism in the brain. This biomarker is measured using the radiotracer fluorodeoxyglucose (FDG) PET or functional magnetic resonance imaging (fMRI). The cerebral blood flow in the brain is associated with neuronal activation which can also be used in the diagnosis of AD; fMRI triggers the changes in blood flow to observe activities in the brain. Sperling et al. (2011) and McKhann et al. (2011) explained the need for more research in clarifying the performance and optimization of the current diagnostic technique, urging for a better way to determine standardized thresholds to distinguish between AD stages. Moreover, current research is trying to find a simpler, cheaper and more efficient way of diagnosing AD,

---

[2]See Section 1.4.1 for more details about neuroimaging techniques in Alzheimer's.

such as through blood tests (Gaugler et al., 2019).

In dementia due to AD, subtle cognitive decline is observable. The cognitive changes can be determined by different ways, such as; 1) obtaining family and medical history in cognitive and behavioural changes, 2) asking family members to provide information about the observed changes in thinking and functionality of the patient, and, 3) cognitive and physical neurologic tests (McKhann et al., 2011). What makes a distinction between MCI (Albert et al., 2011) and dementia due to AD is that, in dementia, the cognitive changes are significant. Meaning that they interfere in the patient's daily activity. The latter is measured by a skilled clinician, who meets the patients and other family members to evaluate the patient's circumstances and daily affairs.

1.2.2 Treatment. So far, no medication can stop or slow down the damage and destruction of brain cells (Gaugler et al., 2019), even though some drugs exist, namely, memantine, donepezil, tacrine, rivastigmine, a combination of memantine and donepezil, tacrine and galantamine. The memantine is effective in preventing the excess stimulation that contributes to damaging neurons, by blocking some receptors in the brain. All remaining drugs help in slowing down the disease symptoms temporarily through increasing the number of neurotransmitters[3] in the brain.

Moreover, non-pharmacologic therapy exists for Alzheimer's dementia that does not require medication. This type of therapy can slow down the cognitive decline, improve the quality of daily life and reduce behavioural symptoms, e.g. aggression, depression and sleep disturbance. Such therapies include computerized training to improve memory, exercising, or changing lighting to reduce sleep disorders. Exercising was shown to reduce the overall cognitive decline in AD (Farina et al., 2014). However, they do not stop or slow the disease progression or the neuron damage.

1.2.3 Brain Changes. There are known biological changes in the brain of an AD patient which affect the neurons by either disrupting their functionality or by killing the neurons. The changes start to take place much earlier than the symptoms arise, that is 18 to 22 years before the onset of symptoms. In other words, the brain is able to compensate for those changes without affecting the patient's cognitive function. However, when more regions in the brain are affected, the patient reaches a point where the brain can no longer compensate for such changes. Hence, a subtle change in cognitive and other symptoms become observable (Gordon et al., 2018).

The healthy human brain consists of roughly 100 billion neurons. They are highly connected through branching extensions - each neuron has its branching extension attached to it. Information is transmitted around the brain using those connections - also called synapses - in the form of

---

[3]Neurotransmitters are endogenous chemicals in the brain with a major role in enabling neurotransmission. They send chemical messages between neurons. They also facilitate communications between a neuron and a muscle or gland cell.

bursts of chemicals released from one neuron to another. Overall, the brain contains 100 trillion synapses. A single neuron has around 7,000 synaptic connections to enhance communication with other neurons. Accordingly, neuronal circuits are formed and facilitate the cellular basis of emotions, memory, vision, movements and other sensations (National Institute on Aging).

The abnormal accumulation of toxic beta-amyloid proteins (called beta-amyloid plaques) between neurons and the abnormal gathering of tau proteins (forming the tau tangles) inside neurons are two main factors characterizing brain change in AD. In a healthy brain, tau stabilizes the microtubules, which are tiny cellular structures that support neurons, and guide nutrients and molecules from the cell body to dendrites. However, in the Alzheimer's brain, abnormal chemicals cause tau to detach from microtubules and form tangles within the neuron, thereby, blocking the nutrient transport inside neurons. Beta-amyloid, on the other hand, has different forms, which result from the break down of amyloid precursor protein (*APP*). The beta-amyloid 42 is especially toxic and may contribute to brain cell death in Alzheimer's. Although ongoing research is still trying to understand how beta-amyloid causes cell death, it appears that cell death happens as a result of the abnormal levels of beta-amyloid protein clumps. This unusual accumulation forms the plaques between neurons and disrupts the cell function.

The tau tangles accumulate at the areas in the brain that are involved in memory. However, when the beta-amyloid plaques reach a certain amount, tau tangles are believed to spread rapidly in neurons throughout the brain. This complex interplay is found to form the basis of AD progression and brain changes (National Institute on Aging). Figure 1.1 explains and visualizes the progress of AD and its different clinical stages. An annotated figure that illustrates the brain regions is show in Figure 1.2.

Some other brain changes that might occur in AD are chronic inflammation and shrinkage (or, atrophy). The inflammation occurs as a result of the activation of microglia. Microglia is an immune cell in the brain activated by tau tangles and beta-amyloid plaques. Microglia cells try to clear the waste caused by dead and dying neurons, or clean the protein collections and plaques. Inflammation happens when microglia is not able to clear all the debris caused. Atrophy, on the other hand, occurs when the loss of neurons as the disease progresses is significant. Specifically, when the brain regions start to shrink (National Institute on Aging).

1.2.4 Risk Factors and Genetic Causes. Though AD is not a normal part of ageing, one of its known risk factors is growing older (Hebert et al., 2010); there is a higher risk of developing AD after the age of 65. Moreover, carriers of the ϵ4 form (or allele) of the apolipoprotein E (*ApoE*) gene are at higher risk of developing the disease compared to non-carriers (Saunders et al., 1993). *ApoE* is the coding gene for a protein responsible for transporting cholesterol in

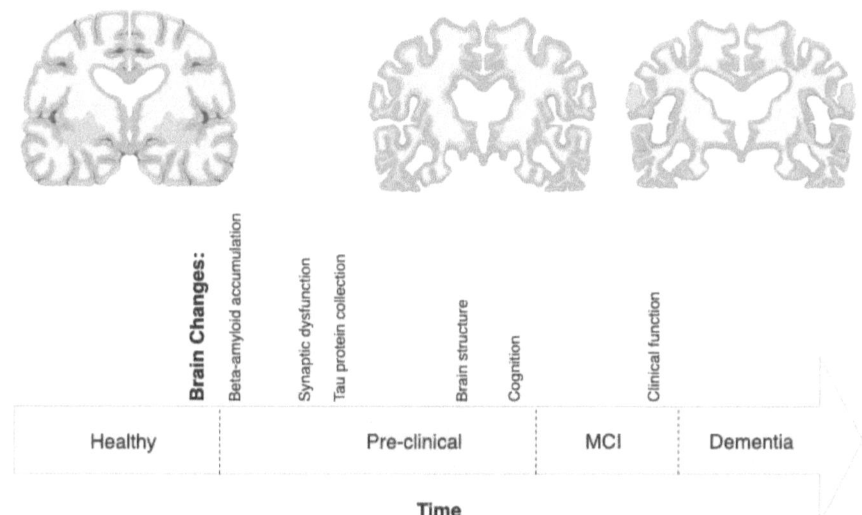

Figure 1.1: Hypothetical and illustrative figure of the different stages in AD; preclinical, MCI and dementia, and the associated changes and biomarkers over time (the order of brain changes from earliest to latest: beta-amyloid accumulation, synaptic dysfunction, tau protein collection, brain structure, cognitive and clinical funtion). The top figures[*] show the amount of change in the brain, as AD progresses, compared to the healthy brain (left figure).

the bloodstream. Family history is another strong risk factor (Fratiglioni et al., 1993); individuals who are members of families with an AD history have a higher likelihood of having the disease. Stroke, high cholesterol, blood pressure, heart disease and head injuries can also contribute to the risk of developing AD and dementia. However, AD is a chronic and complex disorder and is believed to develop as a result of many factors, rather than a single factor, including complex genetic interplay (Van Cauwenberghe et al., 2016).

Known Susceptibility Genes

As per a recent study by Van Cauwenberghe et al. (2016), the *ApoE* gene has three major allelic variants; $\epsilon2$, $\epsilon3$ and $\epsilon4$. These result in six possible *ApoE* genotypes; three in the heterozygous states ($\epsilon2/\epsilon3$, $\epsilon2/\epsilon4$ and $\epsilon3/\epsilon4$), and three homozygous ($\epsilon2/\epsilon2$, $\epsilon3/\epsilon3$ and $\epsilon4/\epsilon4$). Those allelic

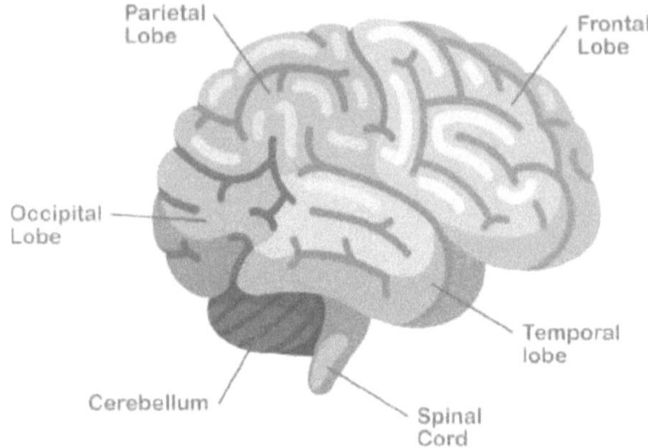

Figure 1.2: An illustrative figure[*] of the main brain regions; the four lobes (frontal, temporal, parietal and occipital) and the cerebellum.

variants play different roles in affecting Alzheimer's; having the $\epsilon4$ in both alleles increases one's risk of developing AD.

ApoE $\epsilon4$ was shown to increase and regulate the onset of cellular deposition of $A\beta$ as the patient ages (Morris et al., 2010). However, this is not sufficient for predicting the disease in asymptomatic individuals. Carriers of two $\epsilon4$ alleles have a 12 fold higher risk of developing AD compared to carriers of two $\epsilon3$ alleles, while carriers of one $\epsilon4$ have 3 times the risk. On the other hand, $\epsilon2$ of *ApoE* has a high protective effect for AD; those who carry $\epsilon2$ allele have a lower risk of developing AD, as shown by Corder et al. (1994).

A small percentage of AD cases, roughly 1%, develop due to mutations[4] in either the *APP*, presenilin 1 (*PSEN1*) or presenilin 2 (*PSEN2*) genes, all of which contribute to $A\beta$ processing (Bekris et al., 2010). If the mutations in *APP* and *PSEN1* are present, there is a 100% chance of developing AD, while the presence of a *PSEN2* mutation implies a 95% chance of developing the disease (Goldman et al., 2011). Other studies (Van Cauwenberghe et al., 2016) have identified genetic variants through genome-wide association studies (GWASs) in some genes as potential risk factors for AD development, such as *SORL1, ABCA7* and *PLD3*. GWAS is the study in which single nucleotide polymorphisms (SNPs) are tested along the genome for associations with a trait (phenotype, or disease) of interest (Hirschhorn and Daly, 2005). A SNP is the smallest form of genetic variation, which occurs at a single locus. However, the biological contribution of these genetic factors to AD progression remains unclear (Bekris et al., 2010).

1.3 Glioblastoma Multiforme

1.3.1 Clinical Characteristics. Glioma is considered to be the most common primary brain tumor, 70 % of which are malignant. It evolves from the supportive cerebral cells and is one of the structural disorders of the central nervous system. The diversity of glioma patients allows the classification of the disease in many ways. Considering the severity of the disease, gliomas can be classified into three categories; diffuse low-grade gliomas, anaplastic glioma and glioblastoma multiforme (Stupp et al., 2010). Although glioblastoma multiforme (GBM; or glioblastoma) is generally not very common; it is most common among adults (average of 55 years), with a mean survival of 15 months.

Epidemiology

GBM accounts for 50% of glioma cases, patients can develop the disease at any age, with a higher incidence between 55 and 60 years (Ohgaki and Kleihues, 2005). Globally, GBM is a rare disease with an incidence of less than 10 per 100,000 individuals. The disease has a poor diagnosis, and has a survival of 14-15 months after diagnosis (Thakkar et al., 2014). Malignant gliomas are responsible for 2.5% of the deaths due to cancer, and are the third highest cause of death due to cancer among patients from 15 to 34 years old (Davis, 2016).

According to the American Cancer Society, the estimated number of brain and other central nervous system cancer new cases is 23,820, with an expected 17,760 deaths in 2019 (Siegel

---

[4]A genetic mutation is an alteration of the nucleotides that make up the sequence of a gene.

et al., 2019). GBM incidence is higher in men than in women with a ratio of 1.26: 1 (Ohgaki and Kleihues, 2005; Thakkar et al., 2014). Incidence of GBM is believed to be higher in the western world, though this could be due to better health reporting of cases in the western world than in developing countries. Some studies report that GBM incidence in Black Africans is less than that in other ethnic groups, e.g. Asians, Caucasians and Latinos. It is especially high in Caucasians resident in industrial areas (Iacob and Dinca, 2009; Ohgaki and Kleihues, 2005).

Symptoms and Diagnoses

Symptoms and presentation of GBM cases vary depending on the associated anatomical structures involved as well as the location and size of the tumor. Patients' complaints might include seizures, headaches, lethargy, nausea, hemiparesis, sight issues, memory problems, or personality changes. The surrounding edema from the tumor might cause more neurological damage than the tumor itself (Young et al., 2015). Some diagnostic strategies include high steroid doses accompanied by gastrointestinal protectant, or dexamethasone 10 mg IV followed by dexamethasone 6 mg every 6 hours (Young et al., 2015), these help to improve the observed symptoms of GBM, and hence, facilitate diagnosis. Patients with seizures can be diagnosed using antiepileptic drugs (AEDs) (Perry et al., 2006). Also, prescription of corticosteroids at the diagnosis stage is useful in controlling vasogenic edema accompanying signs. Other common diagnostic strategies include imaging techniques[5] and testing a sample tissue (biopsy) from the tumor, during or before surgery (using a needle), to determine the cell types of the tumor and the level of aggressiveness.

Classification of Glioma and Glioblastoma

The world health organization (WHO) classification of gliomas uses the level of malignancy, and has four grades (I, II, III and IV). Grade I tumors are normally curable by surgery, and have low-level proliferative activity. Patients with glioma grade II have a low proliferative activity, infiltrative tumor and often recur. Grade III and IV are malignant tumors and invasive. The fourth grade includes glioblastoma, and is the most aggressive type, invasive and fatal (Louis et al., 2007).

Verhaak et al. (2010) utilized a GBM dataset from the Cancer Genome Atlas (TCGA) and analyzed the molecular heterogeneity of GBM. They classified the disease into different molecular subtypes, including proneural, classical, and mesenchymal. These subtypes are based on the expression of platelet derived growth factor receptor alpha/ isocitrate dehydrogenase 1

---

[5]See Section 1.4.2 for more information about imaging techniques in GBM.

(*PDGFRA*)/*IDH1*), the epidermal growth factor receptor (*EGFR*) and neurofibromin 1 (*NF1*), respectively. Besides these three, a fourth subtype described by Verhaak et al. (2010) is neural. These four subtypes respond differently to aggressive treatment protocols (e.g. surgery, radiation or chemotherapy) with the classical being the most responsive and proneural receiving no benefits from the treatment. Accordingly, this suggests that therapeutic strategies should consider the genetic profiles and molecular subtype of tumors.

1.3.2 Treatment. So far, the standard treatment of GBM cases is restricted to surgical resection followed by chemoradiation (a combination of chemotherapy and radiation). The type, size, location of the tumor, and areas affected in the brain are used in planning surgical resection.

Although imaging-guided surgery can reduce the disease symptoms and improve the overall survival, it is not sufficient to cure the disease by itself. Moreover, 80% of GBM cases experience relapse in 2-3 cm around the original lesion (Iacob and Dinca, 2009; Davis, 2016; Hanif et al., 2017), and tumors located in areas like brain stem, eloquent cortex, or basal ganglia are not amenable to resection (Mrugala, 2013). Tumors larger than 6cm show negative effects on survival (Ellor et al., 2014). Additionally, age and other prognostic factors of GBM have been shown to manipulate survival.

Weeks after surgery, a combination of radiotherapy and chemotherapy are used to kill the remaining tumor cells. In particular, the temozolomide, an oral chemotherapy, and stereotactic radiotherapy improve the overall survival (Iacob and Dinca, 2009; Scott et al., 2011; Chinot et al., 2014). Hegi et al. (2005) compared temozolomide effects on GBM patients with and without methylated Methyl Guanine Methyl Transferase (*MGMT*) in tumor cells. *MGMT* is an essential DNA repair protein which protects tumor cells from alkylating chemotherapeutic agents. Iacob and Dinca (2009) found that temozolomide benefits those with a methylated *MGMT* promoter. There are limitations and side effects accompanying radiation therapy, such as permanent neuron damage. This damage is a result of the resistance of some tumors and their invasive nature (Iacob and Dinca, 2009). Temozolomide has its side effects too.

Other types of treatment include clinical trials, in which newly designed drugs with unknown side effects are tried on GBM patients, however, this might be risky (Davis, 2016). Palliative (or supportive) care is also used to improve the patient's quality of life, and relief from pain or other symptoms. Palliative care goes in parallel with other aggressive treatments.

Despite all therapies, these tumors come back for 70% of cases after about a year, and only 5% survive up to 5 years (Thakkar et al., 2014). Most GBM patients succumb to the disease in one year, especially the elderly (Louis et al., 2007). In the case of recurrence, re-resection could be an option for some patients with unclear survival outcomes. Radiation and chemotherapy might also be possible, with some risks accompanying radiation necrosis (Davis, 2016).

1.3.3 Risk Factors and Etiology. The journey of discovering potential risk factors for GBM has not been conclusive, and very little is known so far on this matter. The exposure to ionizing radiation is confirmed to be a physical factor that increases glioma risk (Ellor et al., 2014). Patients usually develop radiation-induced GBM after years of therapeutic radiation from another condition or cancer. Overall, the risk of developing GBM as a result of radiation is 2.5%, with an estimated 116 cases since 1960 (Salvati et al., 2003). Other chemicals and environmental factors include exposure to smoking, vinyl chloride, petroleum refining, pesticides and synthetic rubber manufacturing, however, these factors barely associate with glioma. It is known that less than 1% of glioma patients have a known hereditary disease. Nevertheless, there is an increased risk of developing glioma with some genetic disorders, such as neurofibromatosis 1 and 2, Li-Fraumeni syndrome, retinoblastoma, tuberous sclerosis and Turcot syndrome (Ellor et al., 2014).

Additionally, there are other genetic disturbances associated with cell division and growth in glioma. Examples of such are mutations in genes such as TP53, alpha thalassemia-mental re-tardation syndrome X-linked (*ATRX*), the retinoblastoma 1 (*RB1*), *NF1*, the tumor suppressor phosphatase and tensin homolog (*PTEN*) and telomerase reverse transcriptase (*TERT*) gene. Other genetic factors represent the loss of genetic materials in chromosome 10q, amplification of *EGFR* or alteration in signalling pathways which stimulate cancer cell division and growth, such as the p53 (*TP53*), *PDGFRA*, PI3 Kinase, and Met. Additionally, there are common alterations in tumor suppression pathways including p53, retinoblastoma, and cyclin dependent kinase inhibitor 2A (*CDKN2A*). The primary function of these genes is to prevent uncontrolled growth of cancer cells, however, they are prevented from functioning as a result of the alterations (Nayak et al., 2004; Network et al., 2008; Verhaak et al., 2010). The deregulation of G1/S checkpoints in the cell cycle (YP Lam, 2000) has also been associated with GBM development. Although any combination of the above mentioned factors contributes to the development of GBM, there is high variability within a single tumor, and among different GBM tumors.

1.4   The Science of Neuroimaging

Neuroimaging, or brain imaging, is a rapidly evolving field and newly developed branch in neuro-science, medicine and psychology. It allows imaging of the function or structure of the nervous system. Brain imaging is a non-invasive and very powerful tool that positively contributes to the management of brain-related diseases. Structural and functional imaging are two broad categories of neuroimaging. The main focus areas of structural imaging are, 1) to image the structure of the central nervous system, 2) to diagnose brain diseases and conditions, such as tumors and injuries. Functional brain images provide detailed information about lesions, metabolic diseases, neurological and cognitive psychology (Mabray et al., 2015).

Different types of magnetic resonance imaging (MRI) are commonly used in clinical diagnosis, surgery planning, clinical management, and in following up of brain tumors. They are structural brain imaging that can also visualise functional activities in the brain. MRI is a medical technique which can facilitate disease diagnosis and progression monitoring. It is particularly used in creating images of the physiological process and anatomy of the brain or other organs in the body, using magnetic fields. In most brain diseases and injuries, MRI and other imaging techniques, alongside their measurable characteristics, are adequate in determining the brain structure and function, as well as the anatomical relationships of its different regions.

1.4.1 Neuroimaging Techniques in Alzheimer's Disease. Besides fMRI and PET imaging (see Section 1.2.1), diffusion tensor imaging (DTI) provides further characterization of brain activities and structure. DTI maps the diffusion of water molecules in three dimensions, accounting for the spatial distribution and controlling for the anisotropy (Alexander et al., 2007). More recently, DTI is used in quantifying the connectivity patterns in the brain. This is used in the context of AD, other psychiatric disorders, brain injuries, cancers and a healthy brain. More specifically, DTI techniques are used to reconstruct connectivity information among different parts of the brain in a representation called the connectome (Sporns et al., 2005). The connectome is a network of the brain where nodes represents the different and distinct parts of the brain, and the links are the number of water tracts connecting all pairs of nodes.

1.4.2 Glioblastoma and Brain Tumor Imaging. Different forms of MRI and other types of medical imaging are used to characterize brain tumors in a variety of ways. DTI, fMRI and connectomics are similarly used in tumor imaging (Watanabe et al., 1992; Hart et al., 2015). More specifically, from MRI modalities such as T1, T2 and even the connectome, characteristic features can be determined. MRI imaging modalities help detect the tumor characteristics, it specifically provides information about the shape, size, and location of the tumor in the brain with less radiation exposure. T1 MRI modality provides easy structural annotation of the healthy brain tissues and it makes the tumor border brighter, while the T2 MRI brightens the edema region. Generally, manual segmentation can be carried out by expert radiologists, through segmenting edema from T2 images and Fluid-attenuated Inversion Recovery (FLAIR). The enhancing and non-enhancing structures can be segmented and determined to evaluate specific damage related to the tumor. For example, the VASARI features (Visually Accessible REMBRANDT [Repository for Molecular Brain Neoplasia Data] Images) can be used (Gutman et al., 2013). These features include major and minor axis length of the gross tumor core, the proportion of enhancing and non-enhancing, and the proportion of necrotic tissue (Eisenhauer et al., 2009; Wen et al., 2010; Macdonald et al., 1990). The proportions are estimated using the voxels of the three types of intra-tumoral regions: enhanced, non-enhanced and necrotic, therefore, excluding the edema. Ellingson et al. (2014) reported consensus rules on how to define different types of damage to MRI modalities. More recently, the connectome has also been used to provide further information for surgical planning (Hart et al., 2015).

1.5   Imaging Genetics and Radiogenomics

In AD, and other brain diseases (e.g. psychiatric disorders), imaging genetics is the study of the integrated associations between genetic variants, structural or functional neuroimaging phenotypes and other factors to study the effects of genes on brain connectivity, cognition and behaviour (Thompson et al., 2010; Bedenbender et al., 2011). Imaging genetics has many forms and can take different study designs, and hence, has different dimensionality. It can be conducted as candidate genes association studies with neuroimaging characteristics, or genome-wide variant association studies (Thompson et al., 2010). The imaging part in imaging genetics studies can be a single phenotype (such as hippocampal volume) (Stein et al., 2012) or voxel-wise imaging features. The latter design tests the associations between genetic variants and each voxel in the brain (Stein et al., 2010).

The term radiogenomics, on the other hand, refers to one of two types of studies; 1) the association studies focusing on the response of genetic variations to radiation therapy (Andreassen et al., 2002), 2) the studies which focus on associating gene expression to cancer imaging characteristics, and this design can be used in GBM diagnosis (Diehn et al., 2008). Radiogenomics is used in GBM, and other types of cancer, to identify imaging biomarkers in order to infer the genomics of the disease without the need for biopsy (Chow et al., 2018).

1.6   GWAS Challenges

The past decade has witnessed a significant success of GWAS in discovering thousands of disease-associated variants in a wide range of complex diseases (Visscher et al., 2017). Such discoveries have contributed to improving our understanding of population genetics, the underlying biology of complex diseases, and the discovery of new therapeutics (Visscher et al., 2017). Despite the substantial progress in identifying novel and replicable associations between complex diseases and genetic variants, GWAS has some limitations. One of these is that GWAS mainly studies the association of one SNP at a time, although GWAS successfully identified significant variants, such variants have small effects on the disease, and usually very large sample sizes are required. Consequently, with the contribution of other factors, GWAS only captures a small part of the heritability of complex phenotypes (Manolio et al., 2009). The genetic heritability is defined as the proportion of observed phenotype variation that is due to genotype, this remains small in GWA studies.

Besides small sample size, the above mentioned challenges could be due to lack of complexity in the statistical methods that are used by GWAS. This deficit is caused by either limited knowledge of environmental factors, not involving the complex interplay between other genetic factors, and also considering only the main effects of the SNPs. Recently, Marigorta et al. (2018) highlighted other challenges raised by GWAS, such as the methodological and biological issues, that is, GWAS findings are highly reproducible, however, the ability of these discoveries to predict phenotypes and contribute to precision medicine is limited.

1.7 Project Motivation

In the field of molecular biology, different types of data are increasingly produced to study the human genome variation, and its relationship to other factors and phenotypes. These data are collected and processed to answer different biological questions and to investigate a wide range of hypotheses. Moreover, such data are often collected together with other environmental and clinical information. These consequently led researchers to pose new biological questions and hypotheses, and hence, to develop new methods that aim at incorporating different data types with genetic information.

In the case of brain disorders; brain imaging is usually released along with genetic data. Many approaches have been proposed for integrating both multi-omic and neuroimaging features modelling their association - including, but not limited to, GWA studies. However, they have not overcome the challenges of high dimensionality or the existence of collinearity within and between multiple datasets. Hence, developing new methodologies to address computational issues in brain-related diseases and uncover the genetic architecture of complex diseases is now essential.

Statistics show that patients with AD can become dependent for a long duration before death. This contributes to the public health burden and increases the cases of disability and poor health in the population. Understanding the changes in the brain and their association with other factors in AD contributes to the development and improvement of effective treatments. It can also facilitate early detection of brain changes and enable clinicians to closely observe disease progression.

The genetic heterogeneity in glioblastomas form the basis of treatment specification and clinical outcomes. Obtaining a biopsy to identify the genetic profile of the tumor often takes time in assessment, and might depend on the surgical planning. Radiogenomics is a non-invasive, fast and rapid technique for genetic heterogeneity characterization. The main challenges in treating GBM are the tumor location in the brain and the differences in tumor characteristics between and within patients. Hence, the integration of brain tumor imaging and genetic data are crucial to better understand the complex molecular heterogeneity underlying glioblastoma. This can also facilitate personalized therapy and improve our understanding of disease prognosis.

Genetic association studies have contributed to and improved our understanding of complex brain disease and identified dozens of disease-associated variants in AD and GBM. However, those variants contribute little to the genetic heritabilty of these diseases. In the past decade, imaging genetics and radiogenomics have proven their ability to produce replicatable findings. In this project, we focus on studying the brain structure in connection with the genetics of two brain diseases; AD, which spreads in different areas of the brain, and GBM, which is a focal lesion that affects a particular area in the brain.

Here, we aimed to integrate a wide range of neuroimaging characteristics and multi-omic data, to better understand their associations in the context of brain-related diseases. In particular, we present various association study designs and analysis pipelines in imaging genetics and radio-genomics using datasets provided by TCGA, The Cancer Image Archive (TCIA) and Alzheimer's Neuroimaging Initiative (ADNI). We also propose a robust and unbiased imaging genetics model, using the idea of the structural equation model (SEM). Our model can analyse different scenarios and test different hypotheses of association patterns.

1.8 Project Outline

In this book, we first study a brain disease with local lesions (we consider GBM), then we look at a widespread brain disease which affects the overall brain structure, we consider AD in this part. We test different hypothesis to understand the association between neuroimaging characteristics (e.g. brain connectivity and tumour texture and spatial characteristics) and multi-omic factors (including genome-wide variations and transcriptomic data). We observe the need for a single unified model to study the complex interplay between genetic, environmental and clinical, neuroimaging and phenotype features. We introduce a novel model which can test complex hypotheses in the field of imaging genetics, study the effect of interaction terms on the final phenotype, and accommodate heterogeneous data types including phenotype measurements, brain connectivity, environmental and multi-omic factors. Our model is fast, simple, flexible and assumes no distribution of the data.

This book has four main chapters (Chapter 2 to Chapter 5), besides the Introduction (Chapter 1) and General Discussion (Chapter 6). In Chapter 2, we use Spearman's correlation coefficient test to carry out a multi-stage radiogenomic association analysis, in order to understand the relationship between gene-wide expression and a wide range of spatial and texture features of tumor imaging in a GBM dataset. Chapter 3 and Chapter 4 aim to study the contribution of genetics to the progression of AD utilizing imaging genetics study designs. These chapters propose two longitudinal analysis pipelines to examine the associations between structural connectome changes and genetics factors in AD.

We propose a novel mathematical approach in Chapter 5, which integrates brain imaging, specifically global brain connectivity, genetic and clinical features using the SEM. We apply the proposed method to the same ADNI dataset we used in Chapter 3 and Chapter 4, alongside a simulated dataset. Finally, the General Discussion (Chapter 6), briefly concludes and discusses all the work completed in this book.

Chapter 2 was published as a book chapter in Brainlesion: Glioma, Multiple Sclerosis, Stroke and Traumatic Brain Injuries. Chapter 3 was published in Nature Scientific Reports. Chapter 4 is under review, and Chapter 5 was put in BioRxiv.

Chapter 2

Multi-Stage Association Analysis of Glioblastoma Gene Expression with Texture and Spatial Patterns

2.1 Introduction

Gliomas are the most common type of primary adult brain tumors that arise from glial cells. Gliomas have a very heterogeneous landscape, and they can be classified according to their grade into low-grade glioma, anaplastic glioma, and glioblastoma. The most common and aggressive type of glioma in adults is glioblastoma (GBM), which gives the affected patient an average survival time of only 10 to 18 months. The known molecular classification of GBM into classical, mesenchymal, neural and proneural subtypes is relatively accepted to be related to the expression of *EGFR*, *NF1* and *PDGFRA/IDH1* genes (Verhaak et al., 2010).

Imaging, specifically magnetic resonance imaging (MRI), can offer promising biomarkers reflecting underlying tumor pathology and biological function. If imaging phenotypes of GBM obtained from routine clinical MRI studies can be associated with specific gene expression signatures, quantitative imaging phenotypes will serve as non-invasive surrogates for cancer genomic events and provide valuable information as to the diagnosis, prognosis, and optimal treatment.

Several radiogenomic studies have been carried out for many diseases (Stein et al., 2010; Liu et al., 2009; Batmanghelich et al., 2013; Elsheikh et al., 2018a; Zinn et al., 2011; Gutman et al., 2013; Macyszyn et al., 2015; Binder et al., 2018; Bakas et al., 2017b). For instance for schizophrenia pairs of SNP/Gene and MRI features have been mapped through a linear regression model using PLINK (Stein et al., 2010), genes near significant SNPs were localized. Parallel-ICA, a method that jointly analyse the independent component in multi-modalities through maximising inter-modality correlation and independence,showed promising results (Liu et al., 2009). Batmanghelich et al. (2013) proposed a Bayesian framework to relate imaging and genetic data to phenotypes exploiting connection among these data modalities simultaneously in Alzheimers. Recently, correlations of connectomic features have been related to genes which are known to be related to Alzheimer progression (Elsheikh et al., 2018a). In contrast to Alzheimer's disease and schizophrenia, glioma lesions are generally not spread all over the brain, and local features from MRI can be used. An imaging-genomic analysis study (Zinn et al., 2011) performed by using the tumor volume in T2-weighted FLuid-Attenuated Inversion Recovery (T2-FLAIR) images and large-scale genetic and micro-RNA expression probes demonstrated the potential for molecular subtyping and showed that the high median expression of the *POSTN* gene results in a significant decrease in survival, and for that they used ANOVA and Tukey-Kramer test. Other studies (Gutman et al., 2013; Macyszyn et al., 2015) showed correlations between image feature annotations and expression of genes with glioma molecular subtypes (Verhaak et al., 2010). Specifically, Gutman et al. (2013) found a significant association between contrast-enhanced tumor and these molecular subtypes (Verhaak et al., 2010), where proneural type expressed by *PDGFRA/IDH1* gene showed low levels of contrast enhancement, and the classical type (i.e., primarily described by *EGFR* amplification) correlates with the increased percentage of contrast enhancement. The study used Fisher exact statistics.

Recent population-based studies have assessed the anatomical location of GBM in relation to distinct clinically-relevant molecular characteristics, and have identified the spatial distribution of the tumors being descriptive of their molecular status (Macyszyn et al., 2015; Network, 2015; Ellingson et al., 2012; Ellingson, 2015; Steed et al., 2016; Bilello et al., 2016; Akbari et al., 2018). Furthermore, the emerging research direction of radiomics has shown promise that texture analysis of the various tumor sub-regions in radiographic imaging can also be informative of the tumor's molecular characterization (Aerts, 2016; Lambin et al., 2012; Aerts et al., 2014).

Furthermore, using MRI features for GBM lesions, including texture and shape features, Itakura et al. (2015) proposed a classification imaging method and found three clusters of GBM patients. In their method, they integrate copy number and gene expression data to estimate the molecular pathway activity and show that the three clusters reveal not only different molecular characteristics but also different survival probabilities.

Here, we specifically chose to study the association of gene expression with both the location and texture features of the tumor as the anatomic distribution of gliomas was shown to be heterogeneous within the brain Larjavaara et al. (2007). On the other hand, there are differences in the expression profiles of genes between the invasive and tumor cells in gliomas Hoelzinger et al. (2005). Therefore, studying the genetics underlying the invasion of tumor in association with gene expression might give insights to targeted therapies. The purpose of this chapter is to identify significant associations between gene expression, across the whole genome, and quantitative imaging phenomic features extracted from multi-modal MRI brain scans of patients diagnosed with *de novo* primary GBM. In line with the pre-mentioned studies, here we focus on evaluating the spatial location and texture features of GBM and investigate their associations with gene expression.

2.2 Materials and Methods

2.2.1 Data. For the quantitative association analysis conducted here, we utilized a retrospective cohort of 135 de novo primary GBM patients from the TCGA-GBM collection (Scarpace et al., 2016), with available pre-operative multi-modal MRI scans in The Cancer Imaging Archive (TCIA) (Clark et al., 2013) and corresponding molecular characterization in The Cancer Genome Atlas (TGCA). The multi-modal MRI data we utilized comprise native (T1) and post-contrast

T1-weighted (T1Gd), T2-weighted (T2), and T2-FLAIR modalities. T1 and T2 MRI imaging modalities help to detect the tumor characteristics, they specifically provide information about the shape, size, and location the tumor in the brain with less radiation exposure. T1 provides easy structural annotation of the healthy brain tissues and it makes the tumor border brighter, while the T2 MRI brightens the edema region. The TCGA-GBM subset of 135 patients were identified by Bakas et al. (2017c) as brain scans without any surgically-imposed cavity, and their co-registered and skull-stripped imaging were provided in the TCIA Analysis Results together with expert manually annotated segmentation labels for the various histologically-distinct tumor sub-regions, i.e. enhancing tumor (ET), non-enhancing tumor (NET), peritumoral edematous/invaded tissue (ED) (Figure 2.1) (Bakas et al., 2017c,a). The total sample size of GBM patients reduced to 88 after evaluating patients that had available imaging (Scarpace et al., 2016) and corresponding gene expression. In total, we assessed expression energies for 17815 genes, 11 distinct descriptors of tumor spatial location (Figure 2.2), and 517 radiomic/texture features (Figure 2.2) for each patient's brain tumor scan (Davatzikos et al., 2018; Bakas et al., 2017c,a).

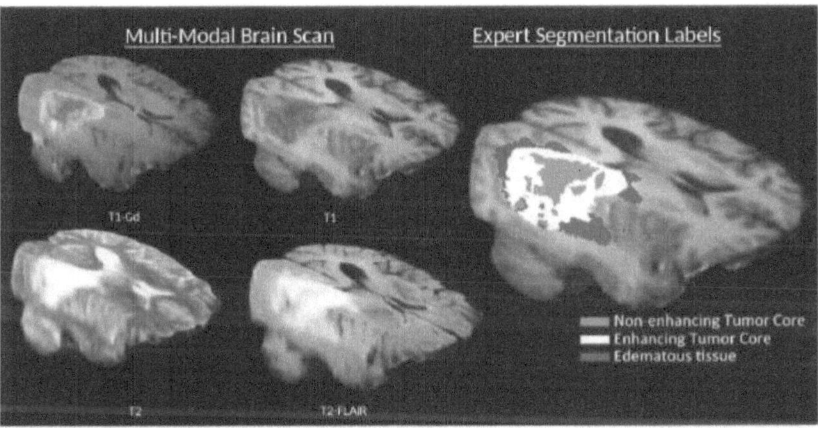

Figure 2.1: Example of a multi-modal MRI brain scan and its corresponding expert segmentation labels.

2.2.2 Quantitative Imaging Phenomic Features.

Radiomic/Texture Features.

We extracted an extensive panel of quantitative texture features, volumetrically (in 3D), for each tumor sub-region as provided by the expert annotations, across all available modalities. Specifically, the texture features we evaluated i) capture global characteristics (i.e., variance, skewness, kurtosis) of each sub-region's intensity distribution on each modality, and ii) include features based on Gray-Level Co-occurrence Matrix (GLCM) which is a matrix that describes the distribution of co-occuring pixel values in gray tone (Haralick et al., 1973) (Figure 2.2), Gray-Level Run-Length Matrix (GLRLM) whose two dimensions are the gray level and run length, each element in the matrix represents the number of pixels with in gray-level for all possible run length values (Galloway, 1974; Chu et al., 1990; Dasarathy and Holder, 1991; Tang, 1998), Gray-Level Size Zone Matrix (GLSZM) which uses 3D brain images to provide information about homogeneous zones in gray matter (Chu et al., 1990; Dasarathy and Holder, 1991; Tang, 1998), and Neighborhood Gray-Tone Difference Matrix (NGTDM) which indicates the gray-tone spatial distribution of voxel intensities (Amadasun and King, 1989).

Spatial Distribution Patterns.

Beyond texture features, we collected discrete spatial information about the anatomical location of each tumor on each brain scan (Figure 2.2). To obtain these spatial distribution patterns we registered all brain tumor scans in a standardized healthy atlas space using an iterative Expectation-Maximization framework (Bakas et al., 2015), while incorporating a biophysical tumor growth model (based on a reaction-diffusion-advection model (Hogea et al., 2008, 2007a,b)) to account for tumor mass effects in the brain parenchyma. We then retrieved the spatial distribution of each tumor according to the discretized anatomical locations of the i) specific lobes (i.e., frontal, temporal, parietal, occipital), ii) insula, iii) basal ganglia, iv) fornix, v) cerebellum, and vi) brain stem. In addition, we also included as distinct features the distances of i) the tumor core (defined as the union of ET and NET), and ii) the ED, from the ventricles.

To produce these quantitative features we have utilized GLISTRboost. Specifically, in the process to produce segmentations of the various tumor sub-regions, the generative part (Gooya et al., 2012) of GLISTRboost, following an Expectation-Maximization framework registers a healthy population probabilistic atlas of glioma patients' brain scans while incorporating a biophysical glioma growth model to account for mass effects. Then, after converting the predicted segmentation in the healthy atlas space, the percentage of the tumor core (i.e., enhancing and non-enhancing tumor) is calculated on each of the brain lobes in this healthy atlas.

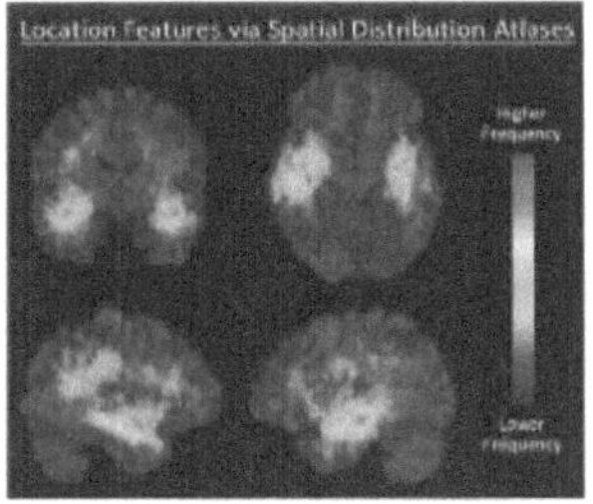
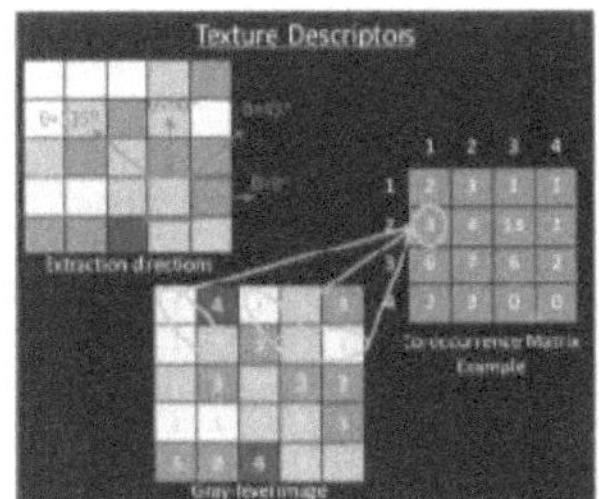

Figure 2.2: Illustrative examples of spatial distribution (left) and texture (right) patterns.

2.2.3 Data Analysis. Initially, we combined the two types of data (imaging - genetics) using the patient ID as a primary column. As a first stage, we used the gene expression and the spatial distribution patterns to perform a non-parametric test of association. To assess the associations, we computed the Spearman correlation coefficient (r_s) between the gene expression, individually, with each of the spatial distribution patterns described in Section 3.2. We then assessed the significance of the correlation coefficient by calculating the p-values as described below.

For each quantitative feature and each gene, we obtained the p-value associated with Spearman correlation coefficient test statistic, which is the p-value of the correlation between a single gene expression with a single feature of the tumor's location in the brain. The Spearman correlation coefficient model for a given feature (y) and given gene expression (x) is;

$$r_s = 1 - \frac{6 \sum_{i=1}^{i=N} d_i^2}{N(N^2 - 1)} \tag{2.2.1}$$

Where d_i is the difference between the ranks of x_i and y_i, and N is 88; representing the number of GBM patients (Kendall et al., 1948). r_s can take any real value between $+1$ and -1; $+1$ represents a strong positive association, -1 means a perfect negative association and 0 indicates no association between the ranks of x and y. Our hypothesis of interest is:

H_0 : *There is no association between the gene expression and the tumor's feature under study*

vs

H_a: *There is an association between the gene expression and the tumor's feature under study,* alternatively:

H_0: $r_s = 0$ *vs* H_0: $r_s \neq 0$

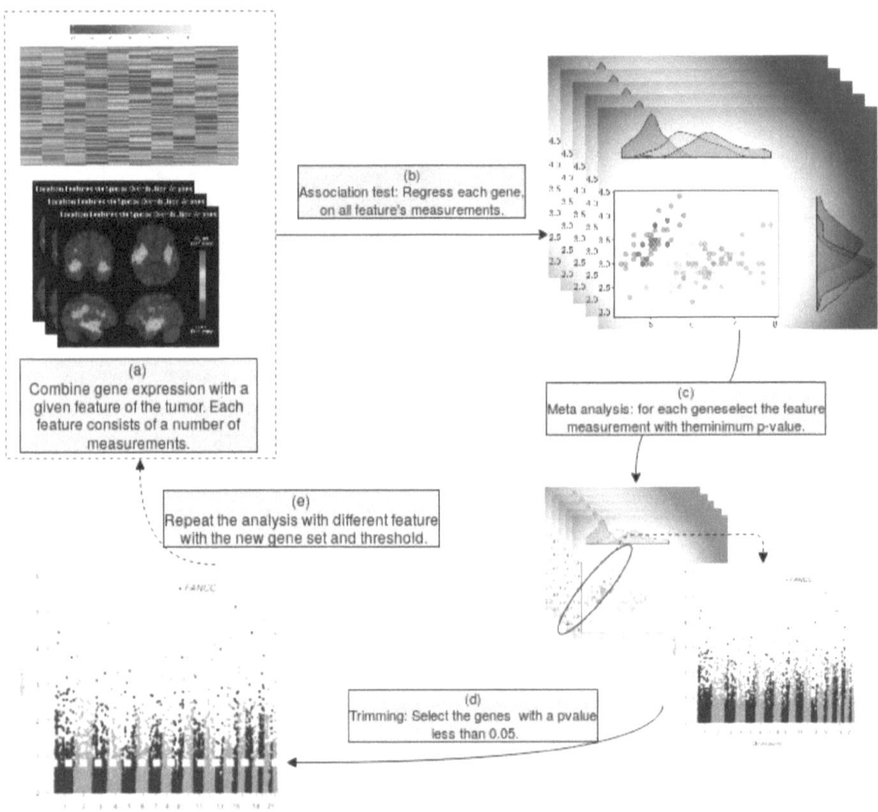

Figure 2.3: Schematic representation of the study's analysis workflow. Step (a) was done using spatial features, while step (e) was done using radiomic features.

To determine the significance of r_s, one can use the t test statistic defined as

$$t_c = r_s \sqrt{\frac{n-2}{1-r_s^2}}, \tag{2.2.2}$$

t_c follows approximately the Student's t distribution with a $N-2$ degrees of freedom under the null hypothesis (Kendall et al., 1948). At a certain significance level, the calculated value of t_c can be compared to the table value obtained from the Student's t distribution (as described previously). The significance of r_s can also be determined using the p-value, which is simply the integration, or the area under the curve from t_c to infinity.

Briefly, in this first stage, the association test was initially conducted to six features of the tumor location (Section 3.2.2). More specifically, for each gene, we computed six p-values, then considered only the minimum p-value at each gene (see Figure 2.3 for the analysis workflow). The latter is referred to as meta-analysis in Figure 2.3 (step(c)). All results reported in Section 2.3 use the summary statistics of the meta-analysis. Moreover, out of all the association results, we excluded all the genes with p-values greater than or equal 0.05. Here we meant to exclude the genes that have very low (and not significant) association with the spatial pattern, which we believe is an important phenotype. This step is referred to as (d) in Figure 2.3. In the second stage, we proceeded with all the genes with p-value less than 0.05, excluding the least significant genes, and we carried the same analysis as in the first stage but using the radiomic features (Section 3.2.2. Table 2.1 shows the thresholds (at both 5% and 10% significance level), along with the number of genes used and remaining in each stage.

Table 2.1: Number of genes, 5% and 10% thresholds used at each stage of the analysis.

Feature					
	No. genes used	Phenotypes	5% threshold	10% threshold	genes after trim
Location	15009	11	0.000000303 ($3.03e^{-7}$)	0.000000606 ($6.06e^{-7}$)	5401
Texture	5401	517	0.000000018 ($1.8e^{-8}$)	0.000000036 ($3.6e^{-8}$)	5370

It is worth mentioning that, out of the total number of genes, we were able to annotate 15009 genes and assign them to their defined physical locations in the DNA. We continued with the first stage of the analysis using those genes (Table 2.1).

2.3   Results

The incidence of tumors specific for region is summarized in Table 2.2. The Manhattan plot for the p-values obtained from the meta-analysis is illustrated in Figure 2.4. The plot shows two horizontal lines which associate with the thresholds of 5% significance level ($\frac{\alpha=0.05}{\text{no. genes}\times\text{phenptypes}}$;top line), and 10% significance level ($\frac{\alpha=0.10}{\text{no.genes}\times\text{phenptypes}}$; bottom line), after correcting for multiple comparisons. The x-axis is the physical position of genes in the DNA, and the y-axis is the negative $log10$ of the p-values. Figure 2.4 also shows the qq-plot of all the genes used in the association analysis. Likewise, each dot corresponds to a p-value of a single gene and $-log10$ of the p-value is used instead. The qq-plot is reported with each Manhattan plot, and it compares the observed distribution of p-values (y-axis) to the expected distribution (x-axis), for each gene tested, where the diagonal line is the null distribution.

Table 2.2: Number and percentage of patients with tumor per brain region

Location	Full name	Number	%
Vent TC	Tumor core (from the ventricles)	88	100.00
Vent ED	Peritumoral edema (from the ventricles)	88	100.00
Frontal	Frontal lobe	63	71.59
Temporal	Temporal lobe	70	79.55
Parietal	Parietal lobe	62	70.45
Basal	Basal ganglia	55	62.50
Insula	Insula	43	48.86
Fornix	Fornix	26	29.55
Occipital	Occipital lobe	35	39.77
Cerebellum	Cerebellum	8	9.09
Brainstem	Brain stem	24	27.27

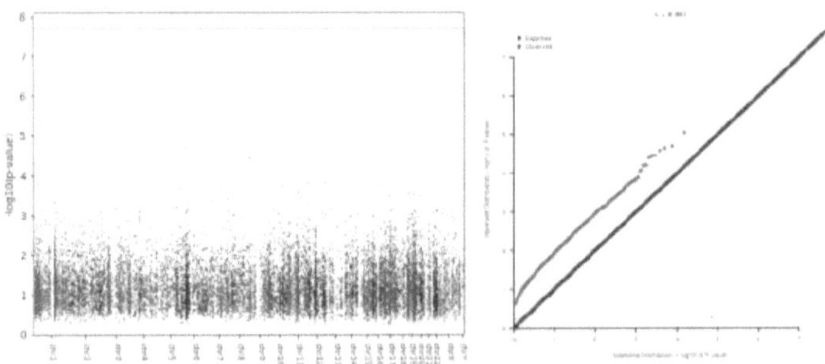

Figure 2.4: A Manhattan (left) and qq-plot (right) of the associations between the tumor spatial distribution patterns, and gene expression. The plot is showing the meta-analysis results.

Table 2.3 shows (only) the top ten p-values and the corresponding genes of the first stage of the analysis. In this stage, none of the p-values was less than $3.03e^{-7}$ or $6.06e^{-7}$ (see table 2.1); therefore, no gene was significantly associated with any of the features. Table 2.3 reports the gene symbol, its start and end position, the associated p-value and feature, and the chromosome.

We then pruned the genes used in the previous stage to a smaller set, by removing the genes that have p-values less than 0.05. With the 5401 genes remaining, we took over the second stage and repeated the same analysis with the texture characteristics of the tumor. The Manhattan and qq-plot for the texture features are shown in Figure 2.5, and Table 2.4 shows the top 10 significant genes. There were no significant hits in this stage as well.

Table 2.3: Top 10 genes: non-parametric association between genes and brain tumor location features in glioblastoma ordered according to the absolute value of r_s.

Gene	Results are sorted according to p-value.					
	start	end	r_s	p-value	Spatial Pattern	Chr
TCN1	59620272	59634048	0.454	8.814e-06	DIST_Vent_TC	chr11
OR2AE1	99473609	99474680	-0.438	2.010e-05	SPATIAL_Basal_G	chr7
KIF13A	17759413	17987854	-0.435	2.271e-05	SPATIAL_Basal_G	chr6
NCBP2	196662272	196669468	0.432	2.619e-05	SPATIAL_Occipital	chr3
RLN2	5299867	5304969	0.426	3.527e-05	SPATIAL_Basal_G	chr9
KCNK9	140613080	140715299	0.426	3.533e-05	SPATIAL_Parietal	chr8
B3GALT6	1167628	1170421	-0.423	3.938e-05	SPATIAL_Brain_stem	chr1
FOXD3	63788729	63790797	0.414	6.0483e-05	SPATIAL_Parietal	chr1
KISS1R	917286	921015	0.414	6.078e-05	SPATIAL_Brain_stem	chr19
PLEKHA8	30067019	30170096	-0.413	6.362-05	SPATIAL_Insula	chr7

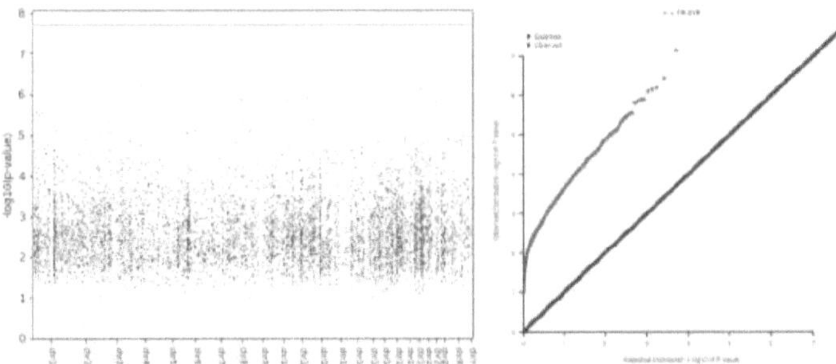

Figure 2.5: A Manhattan (left) and qq-plot (right) of the associations between the tumor texture features, and gene expression. The plot is showing the meta-analysis results.

2.4   Discussion

GBM is a fatal malignant disease that so far is incurable. The identification of genetic risk factors that affect the tumor characteristics improves our understanding of the underlying biological processes for GBM, and may contribute to therapeutic discovery. In this study, we proposed a framework that allows quantifying the non-parametric correlations to test associations between gene expression and different quantitative imaging phenomic characteristics of GBM. Our result has shown a high genetic enrichment through the Manhattan and qq-plots, especially for the texture features (Figure 2.5).

Table 2.4: Top 10 genes: The association of gene expression with texture features of specific GBM sub-regions from specific modalities ordered according to the absolute value of r_s.

Gene	Results are sorted according to p-value						
	Start	End	r_s	p-value	Feature	Region (MRI)	Chr
LRRC46	45908992	45915079	0.537	7.102e-08	GLCM Variance	ED (T2)	chr17
USP38	144106069	144144983	-0.511	3.648e-07	GLSZM SZLGE	ED (T1Gd)	chr4
EPGN	75174189	75181024	0.501	6.542e-07	GLSZM LGZE	ED (T1Gd)	chr4
TUBA1C	49582518	49667114	0.4999	7.096e-07	GLRLM RLV	NET (T1)	chr12
ZNF284	44576296	44593766	-0.498	7.907e-07	GLRLM LGRE	NET (T1Gd)	chr19
IPO8	30781921	30848920	-0.490	1.243e-06	GLRLM GLV	ET (T2)	chr12
MMP7	102391238	102401484	0.490	1.260e-06	GLCM Auto Corr.	ET (T1Gd)	chr11
TLL2	98124362	98273675	0.489	1.342e-06	GLSZM LGZE	NET (T1Gd)	chr10
TRIM55	67039130	67087720	0.488	1.408e-06	GLSZM LGZE	ED (T1Gd)	chr8
UBAP1	34179002	34252521	-0.486	1.582e-06	GLSZM SZLGE	ET (T2)	chr9

Though our results did not highlight genes that significantly associated with the tumor texture features, top hits include *LRRC46, USP38, EPGN, TUBA1C, ZNF284, IPO8, MMP7, TLL2, TRIM55* and *UBAP1* (Table 2.4). *EPGN* expression ($r_s = 0.501$, p-value= $6.542e - 07$) associates with GLSZM LGZE in the T1Gd modality (Table 2.4). *EPGN* was previously reported to be one of the top ten upregulated genes after *EBLN1* silencing in oligodendroglia cells (He et al., 2016). Moreover, the emergence of *EPGN* was identified in another study by Duhem-Tonnelle et al. (2010) in an EGF ligand expression profile, between glioblastoma cell lines and biopsies. Located at chromosome 4, *USP38* ($r_s = -0.511$, p-value= $3.648e - 07$) was associated with GLSZM SZLGE, the modulated *USP38* is known for its involvement in cell growth and stress in the proteasome system (Carminati et al., 2010).

Moreover, as it is illustrated in the Manhattan plot of the spatial features of the tumor (Figure 2.4 and Table 2.3), no gene shows significant association with any of the location features. In addition to the latter, the number of GBM lesions in the cerebellum in clinical settings are quite rare (Drabycz et al., 2010), as also shown in our summary Table 2.2. Our study can provide some insight into this rare type of GBM lesion. Nevertheless, the investigation excluding the patients having those lesions should be repeated in future work.

As the understanding of gliomagenesis grows, several medical imaging biomarkers and genetic variations can be identified, and new hypotheses can be formed. Our proposed genome-wide association framework aims at identifying differentially expressed genes that significantly correlate with various aspects of GBM. The identification of such genes may contribute to the development of targeted therapies that focus on the resistance mechanisms of individual patients.

Through the systematic testing of associations and shrinking of the number of genes at every stage, this pipeline facilitates the evaluation of various hypotheses and reduces the computational complexity. In future work, we plan to extend the study by integrating more quantitative imaging phenomic tumor characteristics, inclusive of morphological, intensity, and volumetric descriptors, as well as parameters derived by biophysical tumor growth modeling.

Chapter 3

Genome-Wide Association Study of Brain Connectivity Changes for Alzheimer's Disease

3.1   Introduction

Alzheimer's disease (AD) is a neurodegenerative disease with believed onset in the hippocampus. It subsequently spreads to the temporal, parietal, and prefrontal cortex (Raj et al., 2015). Symptoms of the disease worsen over time, and as the patient's condition declines, AD ultimately leads to death. Causes of the disease are yet unclear, and it has even been hypothesised to be related to external bacteria (Dominy et al., 2019). However, 70% of AD risk is believed to be contributed by complex genetic risk factors (Ballard et al., 2011). The protein encoded by the apolipoprotein E (*ApoE*) gene, located on chromosome 19, carries cholesterol in the brain, affecting diverse cellular processes. Carriers of the *ApoE* allele $\epsilon 4$ have three times the risk of developing AD compared to non-carriers (Corder et al., 1993). Although *ApoE* $\epsilon 4$ is the primary genetic risk factor that contributes to the development of late-onset AD; its effect accounts for only 27.3% of the overall disease heritability, which is estimated to be 80% (Lambert et al., 2013).

In order to estimate the remaining heritability of AD, many attempts have been made to uncover additional genetic risk factors. Genome-wide studies have successfully identified single nucleotide polymorphisms (SNPs) which affect the development of AD (Bertram and other, 2010; Li et al., 2008; Naj et al., 2010). Understanding the underlying biological process of the disease, and identifying more potential genetic risk variants, could contribute to the development of disease-modifying therapies.

On the other hand, the recent advancements in imaging technologies have provided more opportunities for understanding the complexity of how the brain connects, and at the same time, enhancing and forming a more reliable basis for neuroimaging and human brain research (Mier and Mier, 2015). By merging brain imaging with genetics, previous studies proposed different ways of analyzing the data, to discover genetic factors that affect the structure and function of the human brain. Significant efforts in this area have been made by the Enhancing Neuroimaging Genetics through Meta-Analysis (ENIGMA) (Thompson et al., 2016) project. The methods offered several diverse ways to link together two heterogeneous collections of data - brain imaging and genetic information - depending on the hypothesis under study, and hence, the type of images and genetic information.

Stein et al. (2010) used the T1 weighted Magnetic Resonance Imaging (MRI) scans from the Alzheimer's Disease Neuroimaging Initiative (ADNI), while (Jack Jr et al., 2008) and developed a voxel-based GWAS method (vGWAS) that tests the association of each location in the brain (each voxel), with each SNP. To quantify the phenotype, they used the relative volume difference to a mean template at each voxel, and their method, vGWAS, can be applied to other brain maps with coordinate systems. Although vGWAS did not identify SNPs using a false discovery rate of 0.05, they highlighted some genes for further investigation. More recently, other studies on the genetics of brain structure implemented a genome-wide association of the volume in some sub-cortical regions, and successfully identified significant genetic variants (Stein et al., 2012; Hibar et al., 2015). Additional efforts in the literature include the development of multivariate methods that aim at identifying the imaging-genetics associations through applying sparse canonical correlations to adjust for similarity patterns between and within different clinical stages of the disease (Fang et al., 2016).

Connectomics (Hagmann et al., 2008), or the study of the brain connectome, is a novel advancement in the field of neuroimaging. A structural or functional brain connectome is a representation of the brain, and its connections, as one network. The connectome comprises nodes representing different and distinct regions in the brain, and edges representing the functional or structural connection between brain regions. More specifically, the edges of a structural connectivity network are defined by the anatomical tracts connecting the brain regions (Rubinov and Sporns, 2010). Those are extracted from diffusion weighted imaging (DWI), a type of imaging which detects the diffusion of the water molecules in the brain. Furthermore, the connectome can be summarised by several global and local network metrics (Rubinov and Sporns, 2010) which allow the study of the brain as one entity (one scalar value), the comparison of different groups of participants, as well as the study of variation between and within different brain regions. DWI is not the only method to represent structural connectivity. Structural connectivity can be defined by using T1 scans and voxel-based morphometry (VBM) (Good et al., 2001), a technique investigating structural tissue concentration, especially in gray matter (GM). It has been demonstrated that the morphology across the brain is governed by covariation of gray matter density among different regions (Forsberg et al., 2019). In this way a structural connectome is constructed defining the edges among brain regions as the correlation of GM morphology. This can complement the DWI approach which is mostly based on white matter. Of particular interest are covariability hubs, nodes of this network which have high degree centrality since they are the most representative of the overall cortex (Forsberg et al., 2019; Tijms et al., 2012).

In an attempt to understand aging of the healthy brain, Wu et al. (2013) carried out a longitudinal study of the structural connectome in healthy participants aged 51.1 ± 11.7. Their analysis evaluated the association of the annual change in both the local and global network characteristics with age, but no genetic investigation was carried out in relation to those longitudinal features. However, they found some positive associations between age and connectivity measures at brain regions corresponding to attention, mode and memory. In another study JahanshAD et al. (2013) conducted a GWAS on dementia subjects using connectivity patterns as a phenotype, and identified the genetic variant rs2618516 located in the *SPON1* gene; however, this study considered cross sectional phenotype, collected at one specific time point. VBM-based GWAS have also been carried out combining MCI, AD and control subjects at one specific time point (Shen et al., 2010), identifying the *ApoE* gene and other SNPs related to ephrin receptor as markers strongly associated with multiple brain regions.

In this paper, we used a dataset from ADNI to perform four quanti-tative GWAS, with the longitudinal change in the brain connectome used as a phenotype. Our choice of using the ADNI dataset was because the particular combination of data types needed to run this analysis was available, in the context of AD. We used the absolute difference in the longitudinal integration and segregation global network metrics to represent the change in struc-tural brain connectivity defined by tractography. After obtaining the GWAS summary statistics for all the SNPs typed in the original data, we aggregated their p-values using the PAthway SCor-ing ALgorithm (PASCAL) software (Lamparter et al., 2016) and computed genome-wide gene and pathway-scores. Our result identified a number of genes significantly associated with the change in structural brain connectivity, including *ANTXR2, OR5L1, IGF1, ZDHHC12, ENDOG* and *JAK1*. Most of those genes were previously reported to biologically manipulate the risk and progression of certain neurodegenerative disorders, including Alzheimer's disease (De Ferrari et al., 2007; Kang et al., 2010; Young et al., 2012; Nicolas et al., 2013). Additionally, we investigated whether there are additional changes in connectivity defined by GM covariability.

3.2   Methods

3.2.1 Datasets. Our analysis was conducted on the Alzheimer's Disease Neuroimaging Initiative (ADNI) dataset publicly available . The initiative was launched in 2003 as a public-private partnership, led by Principal Investigator Michael W. Weiner, MD (for updates). To address the aim of our study, we combined two types of ADNI datasets:

1) DWI volumes were taken at two-time points, at the baseline and after 12 months (we refer to this as follow-up). In this set, we used a cohort comprised of 31 Alzheimer's disease patients (age: 76.5 ± 7.4 years), and 49 healthy elderly subjects (77.0 ± 5.1) matched by age, as well as 57 MCI subjects (age: 75.34 ± 5.93).

2) The PLINK binary files (BED/BIM/FAM) genotypic data for AD, controls and Early MCI.

The DWI and T1-weighted were obtained by using a GE Signa scanner 3T (General Electric, Milwaukee, WI, USA). The T1-weighted scans were acquired with voxel size $= 1.2 \times 1.0 \times 1.0$

mm^3 TR = 6.984 ms; TE = 2.848 ms; flip angle=11°). DWI were acquired at voxel size = $1.4 \times 1.4 \times 2.7\ mm^3$, scan time = 9 min, and 46 volumes (5 T2-weighted images with no diffusion sensitization b0 and 41 diffusion-weighted images b=1000 s/mm^2).

3.2.2 Preprocessing of Diffusion Imaging Data. Imaging data have T1 and DWI co-registered. To obtain the connectome; the AAL atlas (Tzourio-Mazoyer et al., 2002) is registered to the T1 volume of reference by using linear registration with 12 degrees of freedom. Despite the fact that the AAL atlas has been criticized for functional connectivity studies (Gordon et al., 2014), it has been useful in providing insights in neuroscience and physiology and it is believed to be sufficient for our case study centered on global metrics. Tractographies for all subjects were generated processing DWI data with the Python library Dipy (Garyfallidis et al., 2014). In particular, the constant solid angle model was used (Aganj et al., 2010), and a deterministic algorithm called Euler Delta Crossings (Garyfallidis et al., 2014) was used stemming from 2,000,000 seed-points and stopping when the fractional anisotropy was smaller than < 0.2. Tracts shorter than 30 mm, or in which a sharp angle (larger than 75°) occurred, were discarded. To construct the connectome, the graph nodes were determined using the 90 regions in the AAL atlas. Specifically, the structural connectome was built as a binary representation when more than 3 connections were given between two regions, for any pair of regions.

3.2.3 Preprocessing for the Gray Matter Analysis. The data for gray matter (GM) analysis were obtained from the T1 volumes of the same subjects. The data have been preprocessed following the optimized VBM protocol from FSL (Douaud et al., 2007). Briefly, volumes have the skull stripped, bias field corrected, then are iteratively registered to a generated template in the MNI space, and have the GM segmented. During the last iteration, data are non-linearly registered to the generated template. The FSL-VBM protocol also introduces a compensation for the contraction/enlargement due to the non-linear component of the transformation: each voxel of each registered grey matter image is multiplied by the Jacobian of the warp field.

3.2.4 Brain Connectivity Metrics. To assess longitudinal changes, we evaluated the following global network metrics at the two time points, at the baseline and follow-up. We then computed the absolute difference between the two measures, at each of the network metrics.

To be in line with previous work on AD and connectomics (PrasAD et al., 2013; Brown et al., 2011; JahanshAD et al., 2013), we focused on specific network segregation and integration features. Segregation represents the ability of a network to form communities/clusters which are well-organized (Deco et al., 2015), while, integration represents the network's ability to propagate information efficiently (Deco et al., 2015).

1) Louvain modularity is a community (cluster) detection method, which iteratively transforms

the network into a set of communities; each consisting of a group of nodes. Louvain modularity uses a two-step modularity optimization (Rubinov and Sporns, 2010). First, the method optimizes the modularity locally and forms communities of nodes, and secondly, it constructs a new network. The nodes of the new network are the communities formed in the previous step. These two steps are repeated iteratively until maximum modularity is obtained, and a hierarchy of communities is formed. For weighted graphs, Louvain modularity is defined as in Equation 4.2.3.

$$Q = \frac{1}{2m} \sum_{ij} \left[A_{ij} - \frac{k_i k_j}{2m} \right] \delta(c_i, c_j), \tag{3.2.1}$$

where A_{ij} is the weight of the edge connecting nodes i and j from the adjacency matrix $\mathbf{A}$, k_i and k_j are the sums of weights of the edges connected to node i and j. respectively, $m = 1/(2A_{ij})$, c_i and c_j are the communities of nodes i and j, and δ is a simple delta function.

2) Transitivity also quantifies the segregation of a network, and is computed at a global network level as the total of all the clustering coefficients around each node in the network. It reflects the overall prevalence of clustered connectivity in a network (Rubinov and Sporns, 2010). Transitivity is mathematically defined by Equation 4.2.4.

$$T^W = \frac{\sum_{i \in N} 2t_i^W}{\sum_{i \in N} k_i(k_i - 1)}, \tag{3.2.2}$$

where t_i^W is the weighted geometric mean of triangles around node i, and k_i its degree.

3) Weighted Global Efficiency is a network integration feature, and represents how effectively the information is exchanged over a network. This feature can be calculated as the inverse of the average weighted shortest path length in the network, as shown in Equation 4.2.1.

$$E^W = \frac{1}{n(n-1)} \sum_{i \in N} \sum_{j \in N, i \neq j} (d_{ij}^W)^{-1}, \tag{3.2.3}$$

d_{ij}^W, is the weighted shortest path length between node i and j, and n is the number of nodes.

4) Characteristic Path Length measures the integrity of the network and how fast and easily the information can flow within the network. The characteristic path length of the network is the average of all the distances between every pair in the network (see Equation 4.2.2).

$$L^W = \frac{1}{n(n-1)} \sum_{i,j \in N, i \neq j} d_{ij}^W.$$

(3.2.4)

where, d_{ij} be the number of links (connections) which represent the shortest path between node i and j.

An illustrative example of global network connectivity metrics is shown in Figure 3.1, the figure consists of a segregated (left) and integration (right) network.

Figure 3.1: An illustrative figure of brain segregation (left) and brain integration (right). In these two figures we have the same nodes and network structure. The brain segregation represents the ability to form sub-networks as the communities on the left figures, while the integration of the brain measures the act of bringing together the different part of the brain as one connected entity, as the thick lines on the right figure.

3.2.5 Gray matter analysis. Connectivity from the GM point of view is defined by the anatomical areas which covary in thickness or volume across the overall brain. Ultimately, the analysis uses another network property given by a hub index described later. Before proceeding with this analysis, a more traditional VBM pipeline was run (Winkler et al., 2014): The method, called "randomise", performs a permutation test for the general linear model. It allows one to compare voxelwise two populations. T-statistics and corrected p-values are then computed. The comparison was carried out within the same populations (AD and control) at the two different time-points, and comparing AD against control subjects.

GM connectivity analysis follows these steps: the GM segmented and registered volumes are further subdivided into cubes of $3 \times 3 \times 3$ voxels which now represent nodes of a network. In this way, each network has on average 6614 nodes. Edges are defined by using the Pearson correlation

r_{jm} computed between two nodes/subvolumes v_j and v_m each time (Tijms et al., 2012):

$$r_{jm} = \frac{\sum_{i=1}^{n}(v_{ij} - \bar{v}_j)(v_{mi} - \bar{v}_m)}{\sqrt{\sum_{i=1}^{n}(v_{ij} - \bar{v}_j)^2}\sqrt{\sum_{i=1}^{n}(v_{mi} - \bar{v}_m)^2}}, \tag{3.2.5}$$

where $\bar{v}_j$, $\bar{v}_m$ are the cubes' mean values, and auto-correlations are set to zero. In the attempt to reduce false positives and with the aim of considering only hubs, once the connectivity matrices are constructed, these are binarized according to a threshold. We set this threshold as 2 standard deviations above the mean, though other more sophisticated threshold choices exist (Tijms et al., 2012; Forsberg et al., 2019). Then, for each node, the degree of connectivity is computed by summing the binarized connections. In this way only the highly connected nodes (hubs) are defined. Lastly, values are averaged first according to the Regions of Interests (ROIs; defined as the structural segmentation of the brain for measuring connectivities) of the AAL atlas, and then for the populations at different time points. Like for the traditional VBM analysis, given the GM hubs defined at two time points we were interested in seeing whether connectivity changes occur within the interval of observations, and whether those are related to the other types of structural connectivity and gene expression.

3.2.6 Integration of the two datasets. To quantify the longitudinal change in brain connectivity, we calculated the absolute difference between the baseline and follow-up for each brain connectivity metric. We then merged the absolute differences with the PLINK fam file, matching the two datasets by the subject ID.

3.2.7 Quality Control.

Quality Control: Individuals

After merging the two datasets into PLINK files, we performed some quality control procedures. First, we applied quality control at individual-level and removed all poor samples, which were identified using PLINK software and the following criteria:

1) Sex-check - here we identified all samples with ambiguous sex and removed them. We used the flag –check-sex .

2) Identifying all the individuals with missing genotype data with the flag –missing. This is to check the missingness rate of genotype information for each individual. In our data, the percentage of missingness for all individuals fell within the range (0.002834, 0.00544), since all subjects passed the threshold of 10% missingness.

3) We then identified Related Subjects (with Identity By Descent (IBD) > 20%), all subjects had IBD between 0.00 and 0.0526. We used a number of PLINK flags, including; –indep-pairwise 50 5 0.2, –extract, –genome, –min and –genome-full

After applying those quality control steps, we had a total of 57 subjects (8 AD, 20 control and 29 MCI) remaining for the rest of the analysis. However, we used a quantitative trait to run the GWAS, that is the change of global connectivity metrics over time.

Quality Control: Genotypes

We ran quality control on the genotypes, by filtering them in terms of their minor allele frequency (MAF) with a threshold of 0.01. All SNPs with less than this threshold are considered rare SNPs and were removed from the analysis. We also removed all SNPs that had missingness more than 33.33 or Genotype Call Rate < 66.67% - this was done in such a way that keeps only SNPs with sample size no less than 38. In addition, SNPs which deviate from the Hardy-Weinberg Equilibrium (HWE) were removed, these are SNPs that have p-values of less than 5e-7 in the HWE test (Wigginton et al., 2005) (in total 351 SNPs did not satisfy the HWE). We used the flags, –maf, –geno and –hwe, and in total 7111195 SNPs remained. Refer to Saykin et al. (2010) for more information about how the genotype data was generated.

Quality Control Correcting for Population Stratification

In this quality control step, we checked for the multiple presence of subpopulations in our sample. This is to make sure if we find significant variants, that the differences in allele frequencies is due to the trait under study and controls for the different ethnic groups. Accounting for population stratification helps to avoid false positives (Hamer and Sirota, 2000). Using multiple ancestry reference genotypic information, we compared the genotypes of each study sample and estimated its ancestry with the Multi-Dimensional Scaling (MDS) analysis (Egs, 2013). We observed that most of our samples belong to the Caucasian population (CEU) and therefore, proceeded by only selecting the Caucasian samples in our study. In Figure 3.2, we show the genotypes of our samples compared with the reference data after the population stratification correction. We included all 57 samples as all belong to the CEU (Caucasian) ancestry.

All previous quality control procedures used here followed the ENIGMA protocol (Egs, 2013). The genetic reference population used here contains 13,479,643 variants that were observed more than once in the European population. These reference data were obtained by ENIGMA from the 1KGP reference set (phase 1 release v3), and imputed.

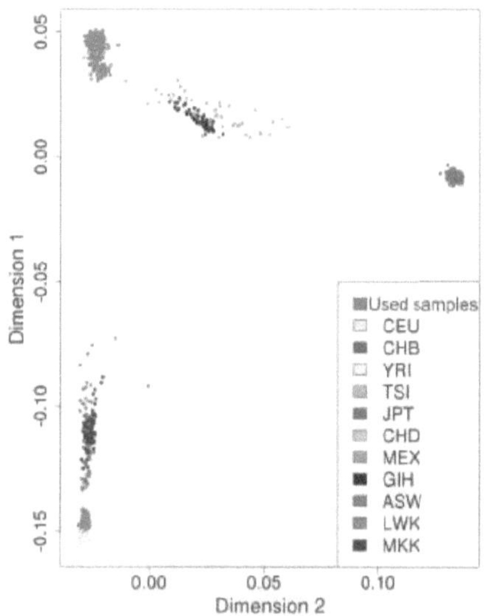

Figure 3.2: Quality control procedures: The plot shows the estimated ancestry of the genotypes of each study sample (in red) after applying the Multi-Dimensional Scaling (MDS). It also compares the genotype of the samples with a multiple ancestry reference. We observed that most of our participants belong to the Caucasian population, denoted here as CEU. A description of the reference population is found in the *Quality Control Correcting for Population Stratification* sub-section.

Quantile Normalisation of Phenotypes

Figure 3.3 and Figure 3.4 indicate that our phenotypes are not symmetrically distributed, and there are potential outliers. Linear models assume symmetric distribution of the response variable. Therefore, to allow the use of linear models and conduct quantitative GWAS for our traits, we first had to normalize our phenotypes. Here, we used PLINK2 (Chang et al., 2015) to perform a quantile normalization (BolstAD et al., 2003) on our phenotypes, using the flag –quantile-normalized.

3.2.8 Integrated Data Analysis.

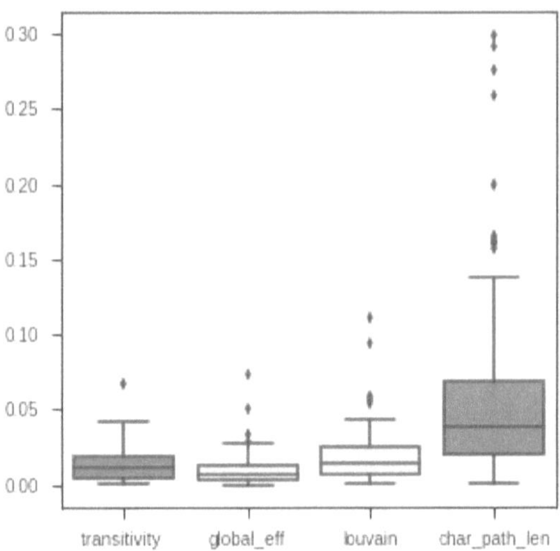

Figure 3.3: Distribution of global network metrics for controls, MCI and AD subjects, combined. Shortcuts stand for; Louvain: Louvain modularity, global_eff: global effeciency, and char_path_len: characteristic path length

Genome-Wide Association Analysis

We performed four quantitative GWAS separately using PLINK software (Purcell et al., 2007) A GWAS for each network connectivity metric measured as the absolute difference between the baseline and follow-up was performed with 57 individuals, and a total of 7111195 SNPs.

Statistical Thresholds

To correct for multiple testing in this analysis, and unless otherwise stated, we rely on the Bonferroni correction (White et al., 2019; Narum, 2006), using the simple equation below:

$$\alpha^* = \frac{\alpha}{M},$$

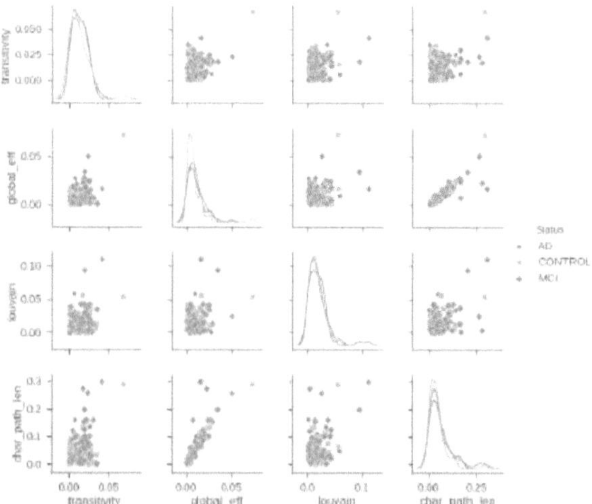

Figure 3.4: Global network metrics scatter plots: The sub-figures show the distribution of the absolute difference of the four network metrics (diagonal plots); as well as the pairwise correlation between them (remaining plots). Each plot compares AD, MCI and controls.

where M is the number of tests of interest (e.g. SNPs, genes or pathways, more information in the results section), and α is the desired significance level. The p-values are then compared with the threshold α^*.

Imputation of GWAS Results

More quality control was done before the imputation of GWAS summary statistics using the Functionally-informed Z-score Imputation (FIZI) Python tool . Using the munge function, 4763 SNPs with duplicated rs numbers and 85757 SNPs with N $<$ 38.0 were removed with a remaining number of 6792416 SNPs. We then imputed the summary statistics with ImpG-Summary - Imputation from summary statistics algorithm (Pasaniuc et al., 2014). In this step, we relied on the European 1000 Genomes (Consortium et al., 2012) haplotypes as a reference panel and performed the Gaussian imputation with FIZI. We managed to impute an additional 2222623 SNPs, all with a maf of < 0.01.

Gene-Wide Scores and Pathway Analysis

After we obtained the GWAS association results, we used them as input for the PASCAL software (Lamparter et al., 2016) to aggregate SNPs at a gene level, and hence, compute gene scores for the four network measures. Along with the obtained association statistics PASCAL uses a reference population from the 1000 Genomes Project to correct for linkage disequilibrium (LD) between SNPs. We set PASCAL to compute the gene score as well as the pathway scores, according to the max of chi-square statistics. We got a p-value for each gene, and for each gene set (or, pathway) provided that there were SNPs present for that gene. Finally, we used Python to plot the Manhattan plot, and R studio (RStudio Team, 2015) to plot the qq-plots. All steps are summarized in the pipeline shown in Figure 3.5.

3.3   Results

3.3.1 Analysis Pipelines. In this work, we used a longitudinal imaging dataset, combined with genetic variation information at the SNP-level. The sample consists of three groups which represent three distinct clinical stages of Alzheimer's disease. This includes healthy individuals (controls), Early mild cognitive impairment (MCI), and Alzheimer's disease. Aiming at studying the genetic effect on the longitudinal change in the brain structure for those groups, we conducted genome-wide tests of the associations between brain image features and different levels of genetic variations. These image features were derived from an intensive map of the brain's neural connections. The overall pipeline followed is summarized in Figure 3.5.

3.3.2 Descriptive Statistics of Brain Imaging Features. Using the DWI images at both baseline and a follow-up visit after 12 months, the brain connectome was constructed. We obtained four global network metrics, as explained in the Methods section. We chose network transitivity and Louvain modularity to represent network segregation, along with characteristic path length and weighted global efficiency to represent the brain integration (Rubinov and Sporns, 2010).

Each of these four metrics quantitatively represents the whole brain network as a single value. Figure 3.6 illustrates both the distribution of the network metrics in the data, in the baseline and follow-up, for the three participant categories. The figure also shows the association patterns of the metrics. A similar figure that illustrates the distribution of the absolute difference between the baseline and follow-up metrics is shown in Figure 3.4, it compares the three groups in each sub-figure. To determine how the differences between these connectivity metrics are distributed,

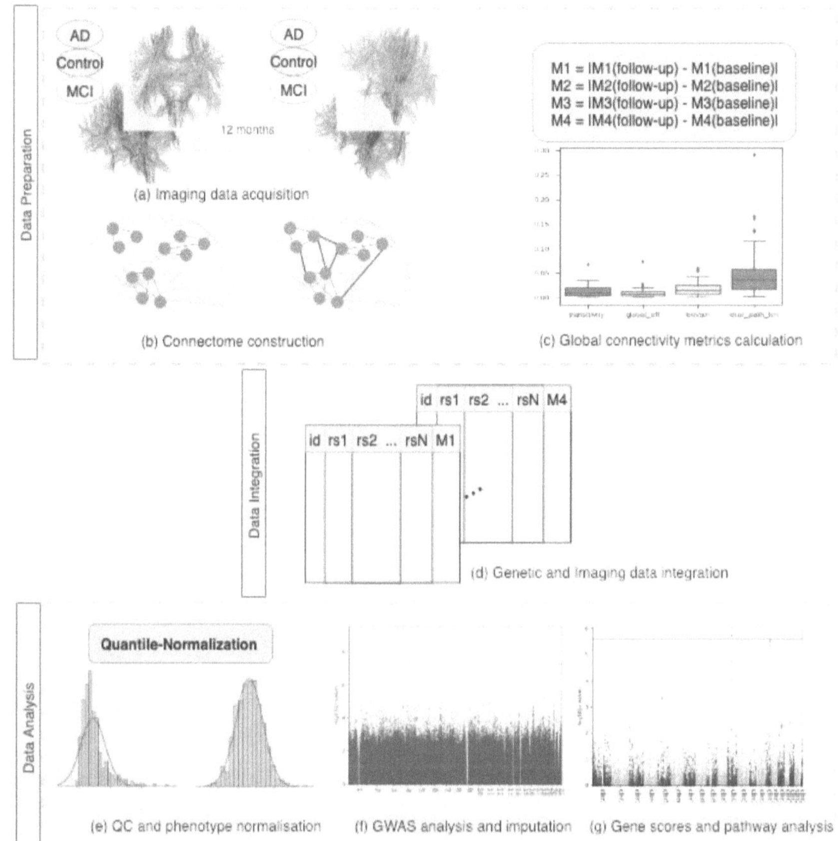

Figure 3.5: The analysis pipeline: (a) The DWI images were collected at two time points, for three clinical stages of AD. (b) The images were processed using distinct brain regions from the Automated Anatomical Labeling (AAL) atlas, and two structural connectomes were constructed for each participant at each time point. (c) Global connectivity metrics were computed, along with the absolute difference between the baseline and follow-up measures. (d) The latter were merged (as phenotypes) with the PLINK FAM files for all subjects present in both datasets. (e) All essential quality control procedures were performed before GWAS analysis, besides the quantile normalization of phenotypes. (f) GWAS was conducted using PLINK, and, (g) the resulting summary statistics were used by PASCAL software to calculate the gene- and pathway-scores accounting for LD patterns using a reference dataset.

we plotted four boxplots, as shown in Figure 3.3, for the three groups combined.

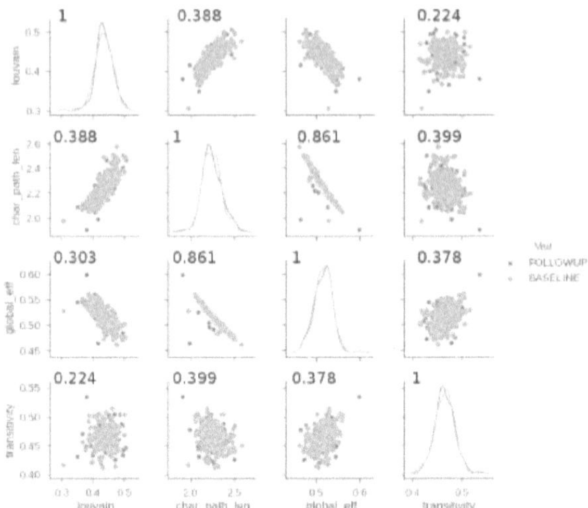

Figure 3.6: Global network metrics scatter plots: The sub-plots compare the four global network metrics before and after 12 months (baseline vs follow-up). Diagonal plots show the distribution of the actual metrics in the baseline and follow-up, while the remaining plots show the correlation between the metrics, for all participants. The numbers next to each sub-figure correspond to the Pearson correlation coefficient.

To verify that the longitudinal change is consistent and not the result of artifacts, we initially compared the imaging features between the two-time points. Table 3.1 shows the results of the non-parametric Wilcoxon test between baseline and follow-up features. The test ranks the values of the paired measurements and compares their central values. In this test, we used the AD patients and controls, most of the metrics turned out to have significant longitudinal differences in the AD brain, but not in a healthy brain. Figure 3.7 compares the four metrics at the two time points, for each group individually, utilizing their boxplots.

We further investigated structural connectivity given by VBM, namely whether there is a structural covariability change in the gray matter. Before doing this, a traditional VBM analysis was carried out. In particular, we compared the AD against the control population at the two time points, and the two populations individually compared to themselves at the two time points. The differences between the groups were given by a t-test converted into corrected p-values (Winkler et al., 2014). Between the same populations at different time points no significant voxels were found, but comparing the two different populations, most of the brain regions were statistically significantly different. Figure 3.8 show the t-statistics map of these differences. In line with previous work (Wang et al., 2018), we identified the peak of statistical difference between AD and control

Table 3.1: Non-parametric Wilcoxon test of the difference between brain connectivity features at baseline and follow-up

Group	**P-values with * are significant (< 0.05)**		
	Network Metric	Statistic	P-value
AD	Characteristic path length	107.0	0.0057*
	global efficiency	98.0	0.0033*
	Transitivity	226.0	0.6664
	Louvain	114.0	0.0086*
MCI	Characteristic path length	612.5	0.0891
	global efficiency	672.0	0.2196
	Transitivity	760.0	0.5972
	Louvain	712.0	0.3630
Control	Characteristic path length	496.0	0.2465
	global efficiency	55.0	0.1172
	Transitivity	517.0	0.3421
	Louvain	529.0	0.4062

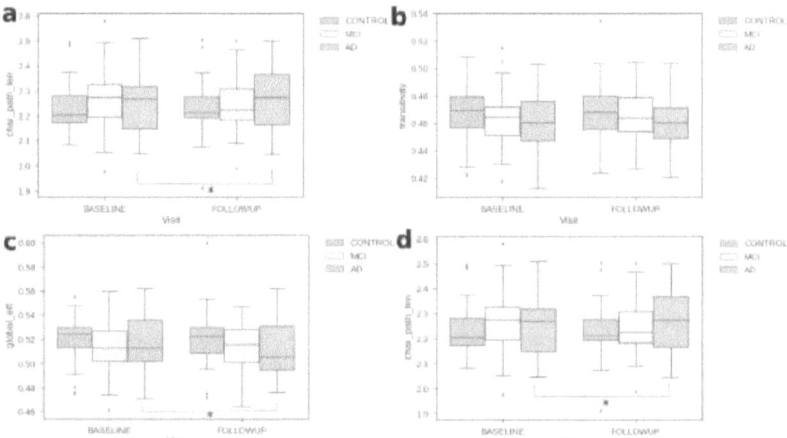

Figure 3.7: Boxplots for global network metrics to compare AD and controls in the baseline (green) and follow-up (yellow). The metrics are, Louvain modularity (a), transitivity (b), global efficiency (c) and characteristic path length (d). It is evident that at least the means for the AD population are different while for the others they are generally unvaried. The asterisk denotes that there is a significant change from baseline to the follow-up visit (p-value< 0.05).

subjects in the hippocampus/parahippocampus, followed by the cingulate cortex and the temporal lobe at both time points. The hubs detection was conducted on the same segmented GM data used in the VBM analysis, and again no significant differences were noted within the same population comparing different time points. The average hubs index is reported in Figure 3.9, and Figure 3.10 depicts the values averaged according to the ROIs of the AAL atlas, specifically showing the hubs index for the AD population at baseline and followup (similar results were obtained for the control population).

Here, the highest values, in line with similar results of previous studies on healthy volunteers, were in the fronto-lateral cortex, cingolum, (Tijms et al., 2012) and basal ganglia (Forsberg et al., 2019). Given the fact that no statistical difference was found for longitudinal changes, both using the traditional VBM analysis and the cortical hubs, no feature of this kind was available for the integrated analysis, which was therefore focused on the structural connectivity given by the tractography.

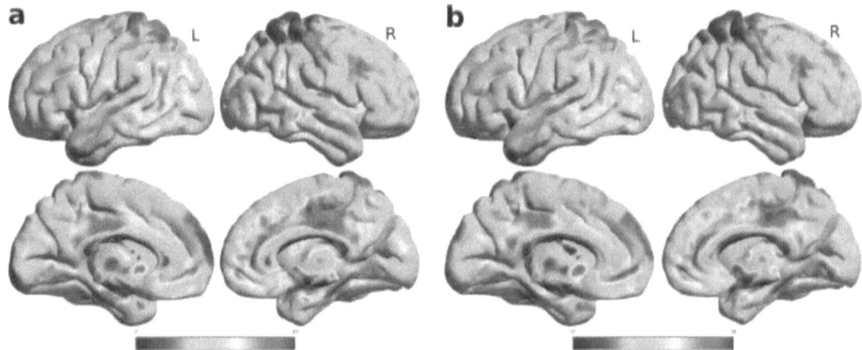

Figure 3.8: T-statistics map of the comparison between the VBM features of AD and control subjects. On the left (a) is the comparison at baseline, and (b) on the right for the followup. All views are for both hemispheres, lateral and medial view. Highest values, depicted in red, were at the hippocampus/parahippocampus, cingulate cortex and temporal lobe for both time points.

3.3.3 Integrated Analysis. After we obtained our phenotypes of interest, given by the longitudinal changes of the features between the two time points, we prepared our data for genome-wide association analysis (see Figure 3.5) by first integrating the phenotypes and genotypes.

The necessary quality control procedures that precede GWAS analysis were run as explained in the Methods section. Briefly, they include cleaning the data such as removing all SNPs with small sample sizes, and individuals with relatedness, as well as population stratification correction. Figure 3.2 shows the plots after correcting for the population stratification. We

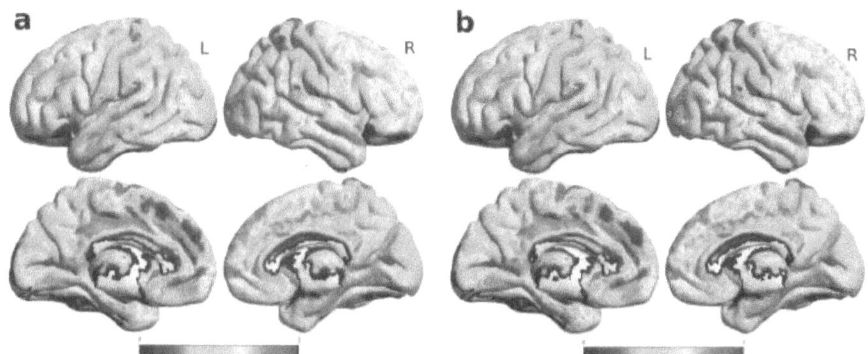

Figure 3.9: Average normalized connectivity hubs, (a) on the left there is the average value at baseline, and (b) on the right for the followup. All views are for both hemispheres, lateral and medial view. Highest values, depicted in red, were at the cingulate cortex, fronto-lateral cortex and basal ganglia, gray areas depict values of 0. The individual values averaged according to the ROIs of the AAL atlas are reported Figure 3.10.

quantile-normalised our phenotypes to allow the use of the linear model in GWAS.

Genome-Wide Association Analysis

GWAS was conducted by regressing the normalised longitudinal changes of global connectivity metrics (response variable) on the SNPs' minor allele frequencies (independent variable), one SNP at a time. Using PLINK (Purcell et al., 2007) we conducted four quantitative GWAS - one for each network metric, after which we performed a Gaussian imputation of GWAS summary results. Figure 3.11 shows the imputed GWAS results for the change in brain segregation metrics. The Manhattan sub-plots appear on the left, while the corresponding quantile-quantile (qq)-plots are on the right. Figure 3.12 shows the imputed GWAS results for brain integration metrics. The x-axis of the Manhattan plot represents the physical location along the genome, while the y-axis is the $(-\log 10(p - value))$, and each dot represents a single SNP. In the qq-plots, the diagonal line represents the expected (under the null hypothesis) distribution of p-values, and similar to the Manhattan plot, each dot in the qq-plot represents a single SNP.

The top 15 SNPs, including the significantly associated SNPs obtained after imputation of GWAS p-values for the absolute difference in Louvain modularity, transitivity, global efficiency and characteristic path length, are shown in Tables 3.2, 3.3, 3.4 and 3.5, respectively. The actual GWAS Manhattan plot for the absolute difference in segregation and integration metrics before imputa-

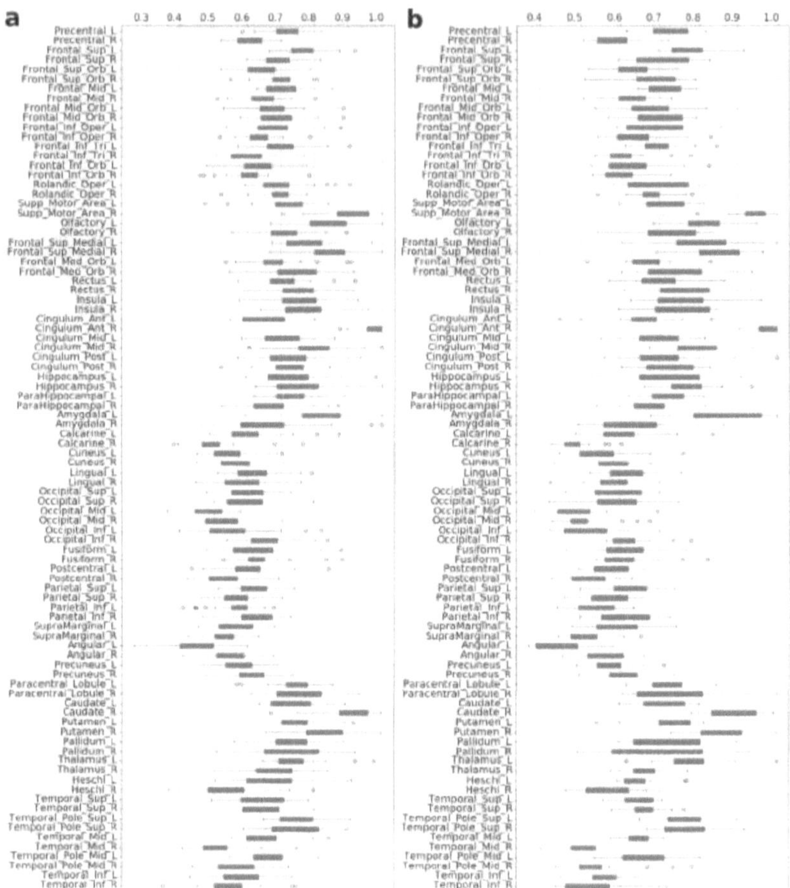

Figure 3.10: Boxplots of hubs degree centrality averaged according to the single ROI of the AAL atlas. On the left (a) are the values for the AD subjects at baseline, and (b) on the right are the values for the AD subjects at follow-up.

tion is provided in Figure 3.13.

Gene and Pathway Scores

Using the imputed GWAS association results (p-values), we computed genome-wide gene scores, along with the pathway (gene set) scores, using the PASCAL software (Lamparter et al., 2016).

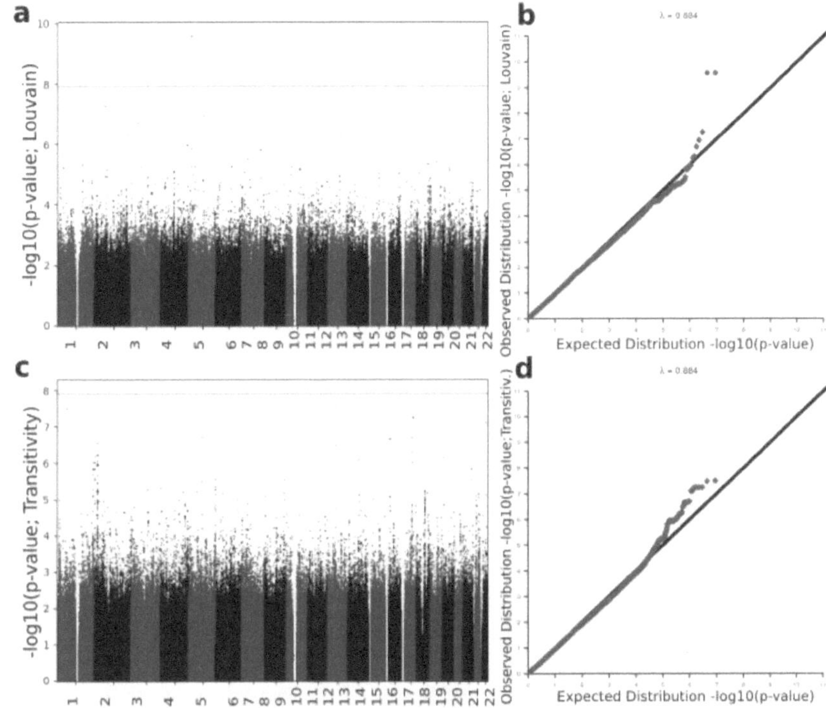

Figure 3.11: Imputation results of GWAS summary statistics for the change in segregation metrics. Top plots represent the change in Louvain modularity phenotype Manhattan plot (a and c) and quantile-quantile (qq)-plot (b and d). Bottom plots represents the change in transitivity phenotype. Louvain modularity imputation results show small evidence of deviation of measures before the tail of the distribution.

Figure 3.14 and Figure 3.15 show the gene scores obtained for brain segregation and integration phenotypes, respectively.

Using the total number of genes in the human genome $(20,000)$ we calculated the threshold. Therefore, we obtain the 5% gene-wide significance threshold by dividing the significance level by the total number of genes (or, tests), i.e.,

$$\frac{0.05}{20,000} = 0.0000025 = 2.5E - 6,$$

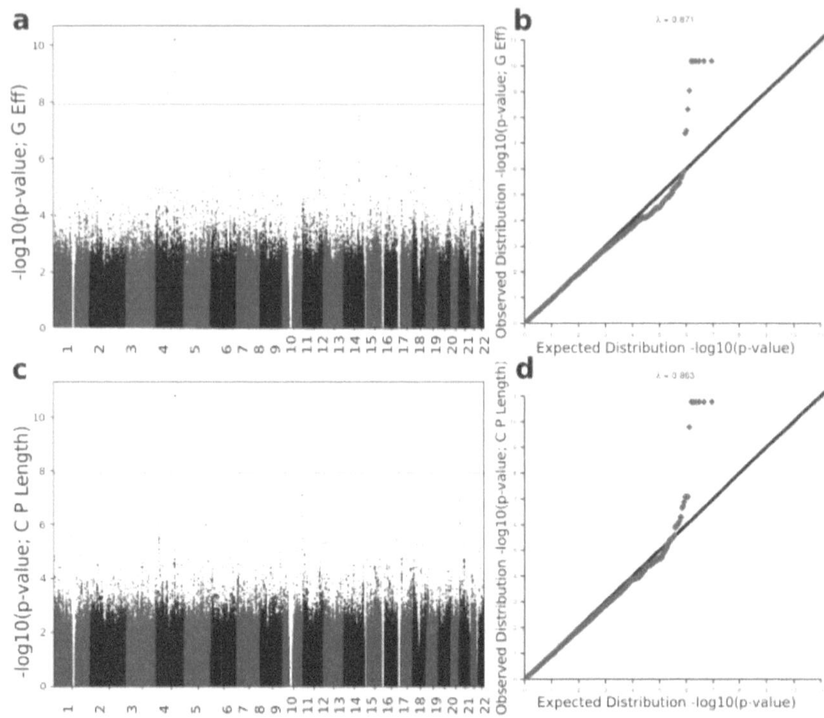

Figure 3.12: Imputation results of GWAS summary statistics for the change in integration phenotypes. Top plots represent the change in global efficiency Manhattan plot (a and c) and qq-plot (b and d), while the plots at the bottom represent the change in characteristic path length phenotype. Both qq-plots show very little evidence of deviation before the tail of the distribution.

If we consider less power (90%) and 10% significance level, we get a gene-wide threshold of

$$\frac{0.10}{20,000} = 0.000005 = 5E - 6.$$

For each gene score result (and for both brain segregation and integration measures) we sorted our results and constructed a table of the top 30 genes (Table 3.6). The table also shows the 5% and 10% significant genes. The gene *CDH18*, contains 3974 SNPs as per the data, was significantly associated with Louvain modularity change over time (p-value $\approx 8.09E - 8$). On chromosome 11 and chromosome 15, a number of genes were associated with the change in brain connectivity through transitivity, while chromosome 9 shows a number of significant association results with characteristic path length.

Table 3.2: Louvain modularity GWAS results: Top 15 SNPs

SNP	Dashed line is the 5% threshold							
	Chr	BP	MAF	Eff/Alt	Type(R2)	Statistic	β	P
rs144596626	5	19473852	0.0079	G/A	imputed (0.717)	6.32 (z)		2.68e-10
rs146631242	5	19396212	0.0079	G/A	imputed (0.717)	6.32 (z)		2.68e-10
rs35942723	2	67943399	0.0066	T/C	imputed (0.7)	5.43 (z)		5.5e-08
rs2460661	5	98903041	0.0211	C/T	imputed (0.65)	5.3 (z)		1.14e-07
rs144454897	3	45220242	0.0065	G/A	imputed (0.839)	-5.2 (z)		2.03e-07
rs12694279	2	213253406		C/A	gwas (1)	-5.992 (t)	-1.19	4.85e-07
rs146293495	11	87983529	0.0066	C/T	imputed (0.782)	-5.01 (z)	0.00923	5.54e-07
rs145955468	4	182593928	0.0066	G/A	imputed (0.818)	4.88 (z)		1.07e-06
rs185097390	4	182653779	0.0065	G/A	imputed (0.818)	4.88 (z)		1.07e-06
rs149021889	5	19168938	0.0079	T/G	imputed (0.603)	4.82 (z)		1.42e-06
rs189745822	7	130949241	0.00791	C/A	imputed (0.805)	-4.8 (z)		1.56e-06
rs193071172	7	130946162	0.00791	G/A	imputed (0.805)	-4.8 (z)		1.56e-06
rs150749209	7	42493974	0.0106	T/C	imputed (0.714)	-4.66 (z)		3.12e-06
rs189358029	17	43702340	0.0092	C/T	imputed (0.437)	4.62 (z)		3.87e-06
rs8053032	16	75181579		G/C	gwas (1)	-5.199 (t)	-0.9548	4.5e-06

Table 3.3: Transitivity GWAS results: Top 15 SNPs

SNP	Dashed line is the 5% threshold							
	Chr	BP	MAF	Eff/Alt	Type(R2)	Statistic	β	P
rs4617614	11	55537046	0.0079	C/T	imputed (0.717)	5.54 (z)		3.11e-08
rs144573130	1	65415375	0.0079	C/T	imputed (0.82)	5.53 (z)		3.27e-08
rs111650215	15	78828083	0.0066	A/G	imputed (0.661)	-5.43 (z)		5.69e-08
rs112671439	15	78819074	0.0066	T/C	imputed (0.661)	-5.43 (z)		5.69e-08
rs113809575	15	78833209	0.0066	G/A	imputed (0.661)	-5.43 (z)		5.69e-08
rs113882269	15	78833286	0.0066	A/G	imputed (0.661)	-5.43 (z)		5.69e-08
rs11912587	22	38371933	0.0171	A/C	imputed (0.738)	5.37 (z)		7.81e-08
rs4459504	15	71883930		G/A	gwas (1)	6.192(t)	0.8938	7.83e-08
rs144750443	5	89847749	0.0066	C/T	imputed (0.745)	-5.2 (z)		1.96e-07
rs3923493	15	71866632		T/C	gwas (1)	5.973 (t)	0.9766	2e-07
rs2518679	14	31252534	0.0065	T/C	imputed (0.679)	5.18 (z)		2.21e-07
rs61156477	14	31266735	0.0145	T/C	imputed (0.679)	5.18 (z)		2.21e-07
rs77762911	14	31255027	0.0145	T/G	imputed (0.679)	5.18 (z)		2.21e-07
rs8018229	14	31261372	0.0145	G/A	imputed (0.679)	5.18 (z)		2.21e-07
rs147801202	2	19102920		A/ATG	gwas (1)	-5.939 (t)	-1.049	2.91e-07

In the pathway results obtained for each metric and each chromosome, the total number of pathways used at each step was 1078. Therefore, the 5% threshold is $0.000046382 = 4.6e - 5$, while the 10% threshold is $0.000092764 = 9.28e - 5$. Table 3.8 reports all the significant results as well as the top 20 pathways along the whole genome and in all the four phenotypes. As shown in the table, REACTOME BIOLOGICAL OXIDATIONS pathway, which consists of genes involved

Table 3.4: Global efficiency GWAS results: Top 15 SNPs

SNP	Dashed line is the 5% threshold							
	Chr	BP	MAF	Eff/Alt	Type(R2)	Statistic	β	P
rs112039371	4	126730783	0.0119	T/C	imputed (0.775)	-6.53 (z)		6.48e-11
rs114045002	4	126746229	0.0119	C/A	imputed (0.775)	-6.53 (z)		6.48e-11
rs76699517	4	126741800	0.0119	T/C	imputed (0.775)	-6.53 (z)		6.48e-11
rs78276525	4	126732179	0.0119	G/T	imputed (0.775)	-6.53 (z)		6.48e-11
rs78538713	4	126742275	0.0119	T/C	imputed (0.775)	-6.53 (z)		6.48e-11
rs78570105	4	126749594	0.0119	G/A	imputed (0.775)	-6.53 (z)		6.48e-11
rs7657714	4	126735951	0.0132	A/C	imputed (0.792)	-6.12 (z)		9.12e-10
rs113323321	4	80897619	0.0132	C/T	imputed (0.743)	5.85 (z)		4.85e-09
rs192963808	12	102764731	0.0066	A/G	imputed (0.694)	5.53 (z)		3.28e-08
rs146655189	12	102674455	0.0079	C/T	imputed (0.717)	5.48 (z)		4.27e-08
rs148061827	10	109846291	0.0066	G/T	imputed (0.678)	4.84 (z)		1.28e-06
rs2139572	12	102767660	0.0119	C/T	imputed (0.657)	4.82 (z)		1.44e-06
rs149903755	14	100808726	0.0079	A/G	imputed (0.804)	-4.77 (z)		1.86e-06
rs149119261	10	108038891	0.0066	C/T	imputed (0.762)	4.73 (z)		2.3e-06
rs62497351	8	17086601		G/T	gwas (1)	5.278 (t)	1.402	2.48e-06

Table 3.5: Charactristic path length GWAS results: Top 15 SNPs

SNP	Dashed line is the 5% threshold							
	Chr	BP	MAF	Eff/Alt	Type(R2)	Statistic	β	P
rs112039371	4	126730783	0.0119	T/C	imputed (0.775)	-6.73 (z)		1.7e-11
rs114045002	4	126746229	0.0119	C/A	imputed (0.775)	-6.73 (z)		1.7e-11
rs76699517	4	126741800	0.0119	T/C	imputed (0.775)	-6.73 (z)		1.7e-11
rs78276525	4	126732179	0.0119	G/T	imputed (0.775)	-6.73 (z)		1.7e-11
rs78538713	4	126742275	0.0119	T/C	imputed (0.775)	-6.73 (z)		1.7e-11
rs78570105	4	126749594	0.0119	G/A	imputed (0.775)	-6.73 (z)		1.7e-11
rs7657714	4	126735951	0.0132	A/C	imputed (0.792)	-6.39 (z)		1.64e-10
rs10113946	9	131534333	0.0211	C/T	imputed (0.725)	-5.36 (z)		8.29e-08
rs11560592	9	131532694	0.0211	C/T	imputed (0.725)	-5.36 (z)		8.29e-08
rs28521006	9	131534909	0.0211	G/A	imputed (0.725)	-5.36 (z)		8.29e-08
rs35354551	20	2112390		C/T	gwas (1)	-6.089 (t)	-1.107	1.31e-07
rs113323321	4	80897619	0.0132	C/T	imputed (0.743)	5.22 (z)		1.83e-07
rs10122433	9	131577388	0.0224	A/C	imputed (0.753)	-5.18 (z)		2.25e-07
rs12236573	9	131555366	0.0224	G/A	imputed (0.715)	-5.03 (z)		5.01e-07
rs10115869	9	131652502	0.0237	G/A	imputed (0.735)	-5.03 (z)		5.03e-07

in oxidation pathways was significantly associated with the change in Louvain modularity at 5% significance level (p-value=2.91E-5), on chromosome 10.

Table 3.6: Top 30 genes: Association results with global network metrics.

Gene	The dashed lines are the 5% ($\frac{0.05}{20,000} = 2.5E - 6$,)and 10% ($\frac{0.10}{20,000} = 5E - 6$.) significant thresholds.				
	Gene IDd	No SNPs	Chromosome	$P - value$	Metric
CDH18	1016	3974	chr5	$8.09359866E - 8$	Louvain
OR5L1	219437	425	chr11	$1.79431531E - 6$	Transitivity
OR5D13	390142	725	chr11	$1.9176502E - 6$	Transitivity
OR5D14	219436	615	chr11	$2.00645586E - 6$	Transitivity
IGF1	3479	446	chr12	$2.05728046E - 6$	G Efficiency
JAK1	3716	670	chr1	$2.58354101E - 6$	Transitivity
PSMA4	5685	306	chr15	$2.74836153E - 6$	Transitivity
AGPHD1	123688	360	chr15	$3.03262918E - 6$	Transitivity
CHRNA5	1138	410	chr15	$3.23116501E - 6$	Transitivity
LOC100506100	100506100	133	chr9	$3.76839713E - 6$	C P Length
ENDOG	2021	272	chr9	$4.17468163E - 6$	C P Length
TBC1D13	54662	283	chr9	$4.18573966E - 6$	C P Length
IREB2	3658	476	chr15	$4.47786255E - 6$	Transitivity
C9orf114	51490	292	chr9	$4.50675833E - 6$	C P Length
ZDHHC12	84885	134	chr9	$4.60713997E - 6$	C P Length
PKN3	29941	174	chr9	$5.47820454E - 6$	C P Length
ZER1	10444	262	chr9	$5.86368468E - 6$	C P Length
LYSMD3	116068	263	chr5	$9.76915765E - 6$	Transitivity
STK35	140901	625	chr20	$1.11193271E - 5$	C P Length
POLR3G	10622	298	chr5	$1.22724816E - 5$	Transitivity
CCBL1	883	394	chr9	$1.42623859E - 5$	C P Length
OR5D18	219438	367	chr11	$1.85630017E - 5$	Transitivity
SCFD1	23256	636	chr14	$2.10167199E - 5$	Transitivity
CDCP1	64866	864	chr3	$2.31367611E - 5$	Louvain
THSD4	79875	2560	chr15	$2.4273E - 5$	Transitivity
MIR548F2	100313771	441	chr2	$2.82926063E - 5$	Louvain
PHYHD1	254295	225	chr9	$3.73413736E - 5$	C P Length
LRRC8A	56262	280	chr9	$4.16333012E - 5$	C P Length
GPR98	84059	1991	chr5	$4.998E - 5$	Transitivity
TMEM200C	645369	338	chr18	$6.73304265E - 5$	Transitivity

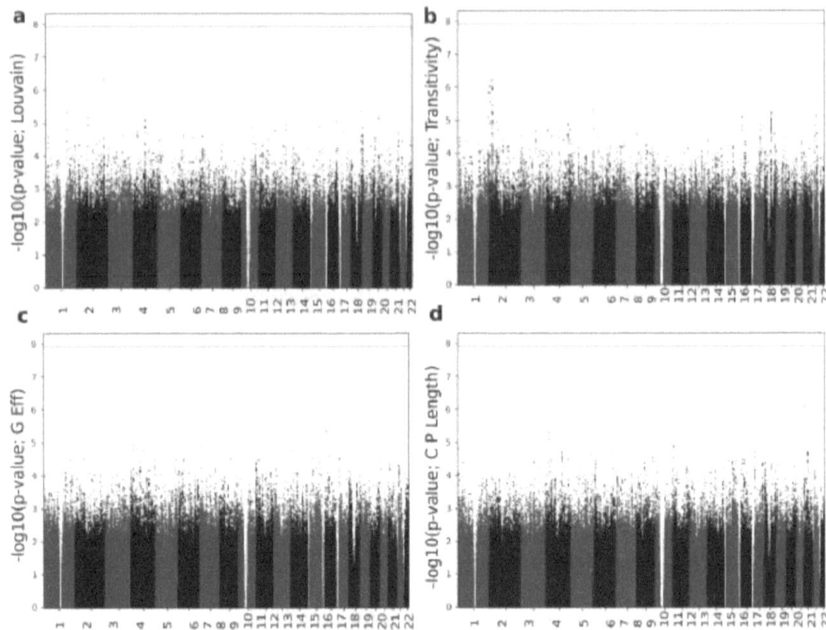

Figure 3.13: Manhattan plots of GWAS results for the change in Louvain modularity (a) and transitivity (b) global efficiency (c) and the change in characteristics path length (d) integration and segregation connectivity metrics.

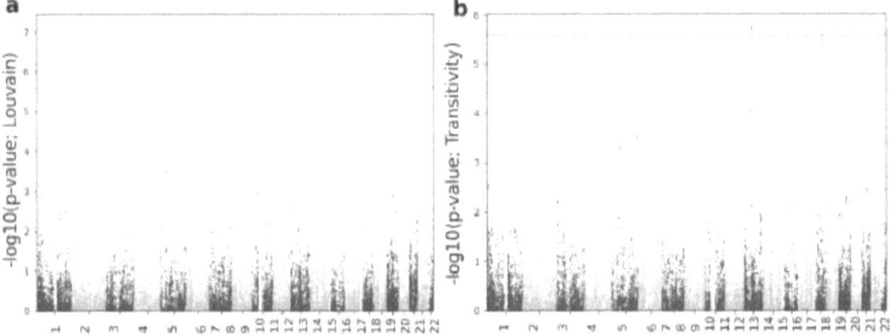

Figure 3.14: Manhattan plots of gene scores derived from imputed summary statistics for the change in segregation metrics. Lovain modularity appears in plot (a), and transitivity is illustrated by plot(b). The horizontal line represents the statistical threshold used here $(2.5E - 6)$.

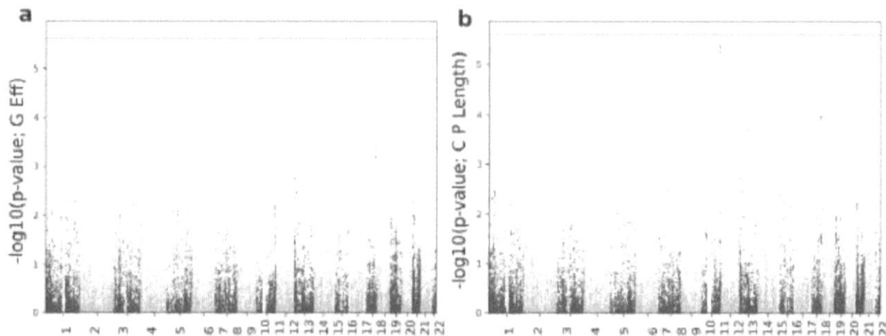

Figure 3.15: Manhattan plots of gene scores derived from imputed summary statistics for the change in integration metrics. Global efficiency is shown in plot (a), and characteristic path length is illustrated by plot (b). The horizontal line represents the statistical threshold used here $(2.5E - 6)$.

Table 3.7: Significant associations between SNPs and global network metrics

SNP ID	Results are sorted according to p-value							
	Chr (Gene)	BP	MAF	Eff/Alt	Type	R2	P	Phenotype
rs144596626	5 (*CDH18*)	19473743	0.0080	G/A	imputed	0.717	2.68e-10	Louvain
rs146631242	5	19396103	0.0080	G/A	imputed	0.717	2.68e-10	Lovain
rs112039371	4	125809628	0.0119	T/C	imputed	0.775	6.48e-11	G Efficiency
rs114045002	4	125825074	0.0119	C/A	imputed	0.775	6.48e-11	G Efficiency
rs76699517	4	125820645	0.0119	T/C	imputed	0.775	6.48e-11	G Efficiency
rs78276525	4	125811024	0.0119	G/T	imputed	0.775	6.48e-11	G Efficiency
rs78538713	4	125821120	0.0119	T/C	imputed	0.775	6.48e-11	G Efficiency
rs78570105	4	125828439	0.0119	G/A	imputed	0.775	6.48e-11	G Efficiency
rs7657714	4	125814796	0.0132	A/C	imputed	0.792	9.12e-10	G Efficiency
rs113323321	4 (*ANTXR2*)	79976465	0.0132	C/T	imputed	0.743	4.85e-09	G Efficiency
rs112039371	4	125809628	0.0119	T/C	imputed	0.775	1.7e-11	C P Length
rs114045002	4	125825074	0.0119	C/A	imputed	0.775	1.7e-11	C P Length
rs76699517	4	125820645	0.0119	T/C	imputed	0.775	1.7e-11	C P Length
rs78276525	4	125811024	0.0119	G/T	imputed	0.775	1.7e-11	C P Length
rs78538713	4	125821120	0.0119	T/C	imputed	0.77	1.7e-11	C P Length
rs78570105	4	125828439	0.0119	G/A	imputed	0.775	1.7e-11	C P Length
rs7657714	4	125814796	0.0132	A/C	imputed	0.792	1.64e-10	C P Length

3.4 Discussion

Association studies of human genome variation and imaging features of the brain have led to new discoveries in AD disease susceptibility. Previous GWAS and Next Generation Sequencing (NGS) identified about 20 genetic loci risk factors associated with AD (Cuyvers and Sleegers, 2016). More recently, cross-sectional studies of GWAS of the brain connectome successfully

Table 3.8: Top 20 gene sets (pathways) results derived from GWAS summary statistics of global network metrics.

Gene Set	Chr	Pvalue	Metric
		Results are sorted by p-value.	
REACTOME_BIOLOGICAL_OXIDATIONS	10	2.91E-5	Louvain
REACTOME_REGULATION_OF_ORNITHINE_DECARBOXYLASE_ODC	15	5.2E-5	Transitivity
KEGG_ALDOSTERONE_REGULATED_SODIUM_REABSORPTION	12	6.6E-5	G Efficiency
BIOCARTA_IL7_PATHWAY	1	6.9E-5	Transitivity
REACTOME_IL_6_SIGNALING	1	7.6E-5	Transitivity
KEGG_PATHWAYS_IN_CANCER	12	1.05E-4	G Efficiency
BIOCARTA_ERYTH_PATHWAY	12	1.32E-4	G Efficiency
BIOCARTA_BAD_PATHWAY	12	1.38E-4	G Efficiency
KEGG_PROGESTERONE_MEDIATED_OOCYTE_MATURATION	12	1.38E-4	G Efficiency
BIOCARTA_IL22BP_PATHWAY	1	1.57E-4	Transitivity
REACTOME_GENERIC_TRANSCRIPTION_PATHWAY	2	2.03E-4	G Efficiency
KEGG_PROSTATE_CANCER	12	2.46E-4	G Efficiency
REACTOME_ACTIVATION_OF_NMDA_RECEPTOR_ UPON_GLUTAMATE_BINDING_AND_POSTSYNAPTIC_EVENTS	8	2.56E-4	Transitivity
REACTOME_POST_NMDA_RECEPTOR_ACTIVATION_EVENTS	8	2.7E-4	Transitivity
KEGG_LYSOSOME	11	3.8E-4	Louvain
REACTOME_OLFACTORY_SIGNALING_PATHWAY	11	4.08E-4	Transitivity
REACTOME_ANTIVIRAL_MECHANISM_BY_IFN_STIMULATED_GENES	1	4.17E-4	Transitivity
REACTOME_REGULATION_OF_IFNA_SIGNALING	1	4.27E-4	Transitivity
BIOCARTA_IL2_PATHWAY	1	4.29E-4	Transitivity
KEGG_GLIOMA	12	4.34E-4	G Efficiency

identified correlations between genetic variants and both AD and dementia (JahanshAD et al., 2013). Incorporating imaging features in a longitudinal setting with genetic information facilitates the identification of additional genetic risk factors which affect AD progression (Elsheikh et al., 2018a). Here, we aim to identify the genetic variations which associate with AD brain neurodegeneration over time. The latter is measured as the change in global network metrics of the brain connectome of three clinical stages of AD.

In this study, we examine the significance of the change in the global network metrics over time, through Wilcoxon test statistics (shown in Table 3.1). We tested the distribution of each metric before and after one year, and only the AD brain showed a difference, compared to controls. We proceeded with the analysis by conducting four quantitative genome-wide association tests, taking the absolute difference in the metrics of brain network integration and segregation as individual phenotypes. To our knowledge, this is the first study of its type, to compare longitudinal imaging features of the connectome to genetic information. These connectivity features were obtained from the structural connectomes defined by tractography. Structural connectomes derived by covariation of cortical morphology was investigated, however, no statistically significant difference at longitudinal level was detected. Despite the belief that covariation of cortical morphology is related to anatomical connectivity of white matter (Pezawas et al., 2004), the technique was not able to detect longitudinal differences in the interval of observation, most likely because these differences are more visible in the "within-brain" connectivity given by tractography, as previously suggested(Forsberg et al., 2019). Therefore, these features cannot be used to perform a GWAS focused on longitudinal changes. Nevertheless, previous GWAS focused on VBM features at one time point (Shen et al., 2010) found an association with the *ApoE* gene and other SNPs related to the ephrin receptor, which are known to be correlated with the loci descried below.

In this data, Louvain modularity analysis identified the SNP rs144596626 (p-value=2.68e-10), in the *CDH18* locus, as the most significant SNP to manipulating changes in brain segregation (See Table 3.2 and Table 3.6). The *CDH18* gene encodes a cadherin that mediates calcium-dependent adhesion, playing an important role in forming the adheren junctions that bind cells. The gene is located on chromosome 5, and it is reported to be highly expressed specifically in the brain, with higher expression in different parts of the Central Nervous System (CNS), including middle temporal gyrus, cerebellum and frontal cortex (Fagerberg et al., 2014). The gene is associated with several neuropsychiatric disorders, as well as glioma, the most common CNV tumor among adults (Bai et al., 2018). Looking at glioma cells, and through *in vitro* and *in vivo* functional experiments, Bai et al. (2018) showed that *CDH18* acts as a tumor suppressor through the downstream gene target *UQCRC2*, and suggested targeting *CDH18* in glioma treatment. Moreover, *CDH18* was reported in a meta-analysis of depression personality trait association as the nearest gene to *rs349475* (Terracciano et al., 2010).

On the other hand, the change in weighted global efficiency metric over time was significantly affected by the *ANTXR2* gene in chromosome 4 (see Table 3.7), through the imputed SNP *rs113323321* (p-value $= 4.85e - 09$) with imputation accuracy of 0.743. *ANTXR2* or ANTXR cell adhesion molecule 2 (also known as *HFS*; *ISH*; *JHF*; *CMG2*; *CMG-2*) is well-known to be involved in the development of Hyaline fibromatosis syndrome (*HFS*) through certain mutations. *HFS* is a collection of rare recessive disorders forming an abnormal growth of hyalinized fibrous tissue; it affects under-skin regions on the scalp, ears, neck, face, hands, and feet. Some studies reported that *ANTXR2* mutations manipulate the normal cell interactions with the extracellular matrix, and its deleterious mutations play an essential role in causing the allelic disorders Juvenile hyaline fibromatosis (*JHF*) and infantile systemic hyalinosis (*ISH*) (Dowling et al., 2003; Hanks et al., 2003). *ANTXR2* interacts with the *LRP6* (Low-Density Lipoprotein receptor-related protein 6) gene, which is located in chromosome 12, and is known for its genetic correlation with *ApoE*. Together, their genetic variants, along with the alteration in Wntβ signalling, might be involved in the development of late-onset AD (De Ferrari et al., 2007).

The segregation of the brain has shown a strong relationship with the olfactory receptor (OR) family 5 (specifically, *OR5L1*, *OR5D13* and *OR5D14* - see Table 3.6), located in chromosome 11, through the change in brain transitivity metric. The OR act together with the odorant molecules in the nose to produce a neuronal response that recognizes smell (O'Leary et al., 2015).

Our findings also suggest that the weighted global efficiency change over time significantly associates with the insulin-like growth factor 1 (*IGF1*) gene, as shown in Table 3.6. A previous study in mouse brain suggests neuroprotection in a mouse model can be obtained through chronic combination therapy with EPO+IGF-I and cooperative activation of phosphatidylinositol 3-kinase/Akt/GSK-3beta signaling. However, they did not test their model in humans (Kang et al., 2010).

At 10% significance level, we identified additional genes associated with Alzheimer's brain segregation and integration alterations. The gene *ZDHHC12* (zinc finger DHHC-type containing 12), as with many others in chromosome 9 - including *LOC100506100*, *ENDOG*, *TBC1D13* and *C9orf114* (p-value$= 3.76839713E - 6$, $4.175E - 6$, $4.186E - 6$ and $4.507E - 6$, respectively) showed a significant score (p-value=4.607E-6) in association with the change in characteristic path length (see Table 3.6). In an in vitro experiment, (Mizumaru et al., 2009) showed that *ZDHHC12* was able to alter amyloid β-protein precursor (*APP*) metabolism, and that the failure of AID/DHHC-12 to regulate the transportation or generation of APP in the neurons might result in the early development of AD (Young et al., 2012). (Singh et al., 2004) also reported the role of Endonuclease G (*ENDOG*) in mediating the pathogenesis of neurotoxicity and striatal neuron death, through exposing the striatal neurons in mouse with Human immunodeficiency virus-1 (HIV-1) Tat_{1-72}.

Located in chromosome 1, Janus kinase 1 (*Jak1*) shows a significant association with the change in transitivity metric. The same phenotype was also reported with other significant gene scores at 10% significance level (Table 3.6) such as the proteasome subunit alpha 4 (*PSMA4*) on chromosome 15, *AGPHD1*, *CHRNA5* and *IREB2* (see Fig 3.14). The dysregulation of the intercellular JAK-STAT signaling pathway, which activates *Jak1* and the Janus kinase protein family, is at the core of neurodegenerative diseases and other brain disorders (Nicolas et al., 2013). *JAK2/STAT3* activation, in particular, was illustrated to protect the neuron, while alteration of the same pathway might play a role in developing neurodegenerative diseases.

We compared our results to previously identified genetic variants in association with Alzheimer's (specifically SNPs), all genetic variants with p-values less than 0.01, in all global network metrics, are summarised in Table 3.9. We retrieved the AD SNP list from Ensembl Biomart online software (Kinsella et al., 2011). Our study reported rs6026398 (β =-0.6496, p-value=0.000814) to be the most significant SNP associated with the change in brain segregation through Louvain modularity. The threshold we set here is $\frac{0.05}{1324}$, as we tested a total of 1324 pathways, though none of the SNPs passed that threshold. Our explanation for this is that variants might play a significant role in developing AD, but do not contribute that much to its progression over time. A way to take this forward is to target all genes known to affect AD susceptibility and test, in a longitudinal study design, which of them contribute to the progression of Alzheimer's disease through the imaging features. Another recommendation is to consider studying the longitudinal association and consider whether any genetic variant has a biased contribution in different brain regions.

One of the main disadvantages of this work is the sample size. We suspect the underestimation that appears in our initial GWAS results for all four phenotypes excluding transitivity, is due to sample size (Figure 3.13). In a larger sample, our result is expected to be more robust and to unveil more variants. However, to some extent, PASCAL (3.14 and 3.15) improved this and unmasked some associations. It is worthwhile mentioning that, in this analysis, we used all the ADNI samples which satisfy our selection criteria. We also considered looking at other datasets (e.g UKBiobank and ENIGMA) but there was no data that matched our specific combination of factors required.

Another concern here, is that our sample size was not sufficient to estimate the genetic correlation and heritability of our phenotypes. Most of the heritability estimation methods requires large sample sizes (at least $\approx$ 5k samples (Bulik-Sullivan et al., 2015; Finucane et al., 2015; Yang et al.,

Table 3.9: The top 22 (p-value < 0.01) association results of AD SNPs obtained from Ensembl BioMart (no one reach the statistical threshold we set $\left(\frac{0.05}{1324}\right)$.

SNP	Results are sorted according to p-value							
	BP	Beta	Statstic	Chr	Eff/Alt	Type(R2)	P-vale	Metric
rs6026398	57180009	-0.6496	-3.544 (t)	20	G/A	gwas (1)	0.000814	Louvain
rs6665019	25328009	0.834	3.466 (t)	1	A/G	gwas (1)	0.00105	Louvain
rs2075650	45395619	0.6423	3.094 (t)	19	G/A	gwas (1)	0.0031	G Efficiency
rs78910009	86408183	NA	-2.9 (z)	16	G/T	imputed (0.83)	0.00368	C P Length
rs11218343	121435587	-1.025	-3.029 (t)	11	C/T	gwas (1)	0.00373	C P Length
rs4746003	71538292	-0.6693	-2.976 (t)	10	T/C	gwas (1)	0.00433	C P Length
rs8014810	36325030	0.7447	2.975 (t)	14	T/G	gwas (1)	0.00434	Transitivity
rs362389	73688861	1.284	2.93 (t)	14	C/A	gwas (1)	0.00492	Louvain
rs73310256	92438849	-1.179	-2.916 (t)	10	C/T	gwas (1)	0.00513	G Efficiency
rs4803760	45333834	-0.5974	-2.868 (t)	19	T/C	gwas (1)	0.00585	Transitivity
rs11218343	121435587	-0.9691	-2.838 (t)	11	C/T	gwas (1)	0.00635	G Efficiency
rs157582	45396219	0.522	2.836 (t)	19	T/C	gwas (1)	0.00641	G Efficiency
rs362384	73686310	0.9341	2.834 (t)	14	A/C	gwas (1)	0.00641	Transitivity
rs58920042	71981089	NA	-2.72 (z)	3	C/T	imputed (0.729)	0.00646	Louvain
rs117780815	124326227	-1.555	-2.793 (t)	6	T/A	gwas (1)	0.0073	Louvain
rs362393	73689629	0.924	2.788 (t)	14	A/G	gwas (1)	0.00732	Transitivity
rs1925690	87867063	0.8101	2.791 (t)	6	T/C	gwas (1)	0.00743	Transitivity
rs7364180	42218856	-0.5574	-2.775 (t)	22	G/A	gwas (1)	0.00754	G Efficiency
rs4746003	71538292	-0.6278	-2.765 (t)	10	T/C	gwas (1)	0.00772	G Efficiency
rs9969729	108631950	-1.04	-2.713 (t)	9	A/G	gwas (1)	0.00911	Louvain
rs117969561	101211189	-1.796	-2.702 (t)	13	T/C	gwas (1)	0.00935	C P Length
rs889555	31122571	-0.4928	-2.689 (t)	16	T/C	gwas (1)	0.00946	G Efficiency

2010)) to yield robust estimates. Besides increasing the sample size, a good practice would be considering more time-points and studying the effect of genes in a survival analysis study design. In this work, we looked at the genetic variations taken at one time-point, and converted the longitudinal imaging information into a single measurement to study their association. A possible future focus would be to incorporate clinical and environmental factors such as hypertension and dementia score as well as the gene-gene and gene-environment interactions.

In summary, we conducted a longitudinal study and proposed a fast and straightforward way to quantify changes in the brain connectome through global connectivity measures of 1) segregation, through Louvain modularity and transitivity, and 2) integration. For the latter, we used two metrics including the characteristic path length and the weighted global efficiency. We conducted a genome-wide analysis, starting with four quantitative GWAS, regressing the pre-mentioned global network metric on all SNPs, and then computed the gene scores by aggregating the GWAS summary statistics at a gene-wide level. In the ADNI sample we used here, and at a power of 95%, despite the small sample size we identified significant SNPs and genes. The Louvain modularity change was affected by the *ANTXR2* gene, while through transitivity, the change in brain connectivity is associated with *OR5L1*, *OR5D13* and *OR5LD14*. On the other hand, the integration of the brain is affected by *IGF1*. In previous studies, connectome changes in AD have been identified, moreover, connectome genetics studies attempt to identify the association between brain connectivity features with genetics. Results in this area are often limited to isolated brain areas, or global connectivity measures, and in both cases, findings lack the understanding of the molecular consequences, causes of the connectome changes, and its contribution to drug development in AD. A greater understanding of the genetic contribution and relationship of these genes and their effect over time through targeted studies, might facilitate the development of drug therapy to reduce the disease progression.

Chapter 4

Relating Global and Local Connectome Changes to Dementia and Targeted Gene Expression in Alzheimer's Disease

4.1 Introduction

There are many factors which may affect the susceptibility to Alzheimer's Disease (AD) and various ways to measure the disease status. However, there is no single factor which can be used to predict the disease risk sufficiently (Barnes and Yaffe, 2011). Genetics is believed to be the most common risk factor in AD development (Gatz et al., 1997). Towards studying the etiology of the disease, a number of genetic variants located in about 20 genes have been reported to affect the disease through many cell-type specific biological functions (Gaiteri et al., 2016). Those efforts resulted from omic studies such as Genome-Wide Associations Studies (GWAS). GWAS highlighted dozens of multi-scale genetic variations associated with AD risk (Lambert et al., 2013; Escott-Price et al., 2014; Elsheikh et al., 2020b).

From the early stages of studying the disease, the well known genetic risk factors of AD were found to lie within the coding genes of proteins involved in amyloid-β($A\beta$) processing. These include the well-known Apolipoprotein E (*ApoE*) gene that increases the risk of developing AD (Corder et al., 1993), the Amyloid precursor protein (APP) (Goate et al., 1991), presenilin-1 (*PSEN1*) and presenilin-2 (*PSEN2*) (Levy-Lahad et al., 1995; Rogaev et al., 1995). More recently, the advancement in technologies and integration of genetic and neuroimaging datasets has taken Alzheimer's research steps further, and produced detailed descriptions of molecular and brain aspects. Such studies have shown a great success in unveiling and replicating previous findings (Medland et al., 2014; Elsheikh et al., 2018b). Shaw et al. (2007), for example, showed that carriers of ApoE are more likely to lose brain tissue, measured as the cortical gray matter, than non-carriers. Other studies have utilised the *connectome* (Hagmann et al., 2008) to study different brain diseases through associating genetic variants to brain connectivity (Thompson et al., 2014). A structural *connectome* is a representation of the brain as a network of distinct brain regions (nodes) and their structural connections (edges), calculated as the number of anatomical tracts. Those anatomical tracts are generally obtained by diffusion tensor imaging (DTI) (Alexander et al., 2007), a method used for mapping and characterizing the diffusion of water molecules, in three-dimensions, as a function of the location. This representation highlighted a network based organization of the brain with separated subnetworks (*network segregation*) which are connected by a small number of edges (*network integration*) (Deco et al., 2015). Given such a "small-world" representation of the brain, it is also possible to represent each individual brain as single scalar metrics which summarize peculiar properties of segregation and integration (Rubinov and Sporns, 2010). Alternatively, those global metrics can also be used to quantify local properties of specific nodes/areas. Early works demostrated that *ApoE-4* carriers have an accelerated age-related loss of global brain interconnectivity in AD subjects (Brown et al., 2011), and topological alterations of both structural and functional brain networks are present even in healthy subjects carrying the *ApoE* gene (Chen et al., 2015). A more recent study has shown association between ApoE expression and brain segregation changes (Elsheikh et al., 2018a). Going beyond the *ApoE* gene, JahanshAD et al. (2013) used a dataset from Alzheimer's Disease Neuroimaging Initiative (ADNI) to carry out a GWAS of brain connectivity measures and found an associated variant in F-spondin (*SPON1*), previously known to be associated with dementia severity. A meta-analysis study also showed the impact of *ApoE*, phosphatidylinositol binding clathrin assembly protein (*PICALM*), clusterin (*CLU*), and bridging integrator 1 (*BIN1*) gene expression on resting state functional connectivity in AD patients (Chiesa et al., 2017).

Moreover, AD is a common dementia-related illness; in the elderly, AD represents the most progressive and common form of dementia. Accordingly, incorporating and assessing dementia severity when studying AD provides more insights about the disease progression from a clinical point of view. A reliable global rating of dementia severity is the Clinical Dementia Rating (CDR) (Morris et al., 1997). This paper uses a dataset from ADNI and presents an integrated association study of specific AD risk genes, dementia scores and structural connectome characteristics. Here, we adapted a longitudinal case-control study design to mainly examine the association of known AD risk gene expression with local and global connectivity metrics. We also aim at testing the longitudinal effect of brain connectivity on different CDR scores, and carrying out a multivariate analysis to study the longitudinal effect of gene expression and connectome changes on CDR. Our approach can be summarized in the simplistic represen-tation in Figure 4.1, where specific genes affect decreases in connectivity comparing baseline and follow-up and this ultimately affects intellectual abilities and CDR scores. Although it is more useful to extract the gene expression profiles from the brain, we used the gene expression from blood samples in this work as provided by ADNI. Blood gives a general idea of what is happening in the body, and can detect differences in gene expression. Moreover, blood samples are easy to obtain and are noninvasive.

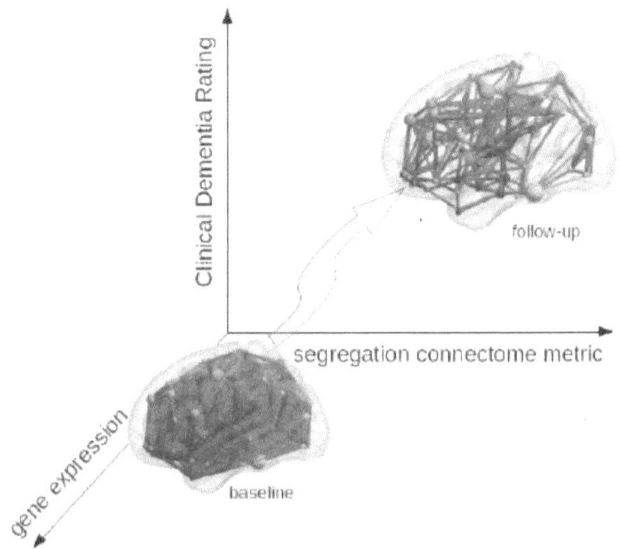

Figure 4.1: Simplistic representation of our approach which relates connectome metrics of segregation (disconnection), cognitive decline and gene expression.

4.2 Materials and Methods

4.2.1 Data Description. We used two sets of data from ADNI, which is available at . To fulfil our objectives, we merged neuroimaging, gene expression and CDR datasets for all the participants with those three types of data at two time points available. We considered follow-up imaging and CDR acquisition one year later than the baseline visit. Given those constraints, we ended up with a total of 47 participants. We adopted a case-control study design; 11 of the participants are AD patients, while 36 are controls. The data were matched by age, and the distribution of age in AD ranges between 76.5±7.4 for cases, and 77.0±5.1 years in controls.

Imaging Data

For the imaging, we obtained the DTI volumes at two time points, the baseline and follow-up visits, with one year in between. Along with the DTI, we used the T1-weighted images and they were acquired using a GE Signa scanner 3T (General Electric, Milwaukee, WI, USA). The T1-weighted scans were obtained with voxel size $= 1.2 \times 1.0 \times 1.0 mm3 TR = 6.984 ms$; TE $=$ 2.848 ms; flip angle$= 11°$), while DTI obtained with voxel size $= 1.4 \times 1.4 \times 2.7 mm3$, scan time $= 9$ min, and 46 volumes (5 T2-weighted images with no diffusion sensitization b0 and 41 diffusion-weighted images b$= 1000s/mm^2$).

Genetic Data Acquisition

We used the Affymetrix Human Genome U219 Array profiled expression dataset from ADNI. The RNA was obtained from blood samples and normalised before hybridization to the array plates. Partek Genomic Suite 6.6 and Affymetrix Expression Console were used to check the quality of expression and hybridization (Saykin et al., 2015). The expression values were normalised using the Robust Multi-chip Average (Irizarry et al., 2003), after which the probe sets were mapped according to the human genome (hg19). Further quality control steps were performed by checking the gender using specific gene expression, and predicting the Single Nucleotide Polymorphisms from the expression data (Vawter et al., 2004; Schadt et al., 2012)

In this work, we targeted specific genes which have been reported to affect the susceptibility of AD. We used the BioMart software from Ensembl to choose those genes by specifying the phenotype as AD (Smedley et al., 2015). We obtained a total of 17 unique gene names and retrieved a total of 65 probe sets from the genetic dataset we are using here.

Clinical Dementia Rating

The Clinical Dementia Rating, or CDR score is an ordinal scale used to rate the condition of dementia symptoms. It range from 0 to 3, and is defined by five values: 0, 0.5, 1, 2 and 3, ordered by severity, which stand for none, very mild, mild, moderate and severe, respectively. The scores evaluate the cognitive state and functionality of participants. Here, we used the main six scores of CDR; memory, orientation, judgement and problem solving, community affairs, home and hobbies, and personal care. Besides the latter, we used a global score, calculated as the sum of the six scores. We obtained the CDR scores at two time points in accordance with the connectivity metrics time points.

Connectome Construction

Each DTI and T1 volume have been pre-processed performing Eddy current correction and skull stripping. Given the fact that DTI and T1 volumes were already co-registered, the AAL atlas (Tzourio-Mazoyer et al., 2002), and the T1 reference volume are linearly registered according to 12 degrees of freedom. Tractography for all subjects has been generated by processing the DTI data with a deterministic Euler approach (Garyfallidis et al., 2014), using 2,000,000 seed-points and stopping when fractional anisotropy (FA) is smaller than 0.1 or a sharp angle (larger than $75°$). To construct the connectome, we assigned a binary representation in the form of a matrix whenever more than three connections were present between two regions of the AAL, for any pair of regions. Tracts shorter than 30 mm were discarded. The FA threshold was chosen in a such a way that allows reasonable values of characteristic path length for the given atlas. Though the AAL atlas has been criticized for functional connectivity studies (Gordon et al., 2014), it has been useful in providing insights in neuroscience and physiology, and is believed to be sufficient for our case study (Gordon et al., 2014).

4.2.2 Global and Local Connectivity Metrics. To quantify the overall efficiency and integrity of the brain, we extracted global measures of connectivity from the connectome, represented here in four values of network integration and segregation. Specifically, we used two network integration metrics 1) the global efficiency (E; Equation 4.2.1), and 2) the weighted characteristic path length (L; Equation 4.2.2). Both are used to measure the efficiency of which information is circulated in a network. On the other hand, we used; 1) Louvain modularity (Q; Equation 4.2.3), and 2) transitivity (T; Equation 4.2.4) to measure the segregation of the brain, that is, the capability of the network to shape sub-communities which are loosely connected to one another while forming a densely connected sub-network within communities (Deco et al., 2015; Rubinov and Sporns, 2010).

Suppose that n is the number of nodes in the network, N is the set of all nodes, the link (i, j) connects node i with node j and a_{ij} define the connection status between node i and j, such that $a_{ij} = 1$ if the link (i, j) exist, and $a_{ij} = 0$ otherwise. We define the global connectivity metrics as;

$$E = \frac{1}{n(n-1)} \sum_{i \in N} \sum_{j \in N, j \neq i} d_{ij}^{-1}, \tag{4.2.1}$$

where, $d_{ij} = \sum_{a_{uv} \in g_{i \leftrightarrow j}} a_{uv}$, is the shortest path length between node i and j, and $g_{i \leftrightarrow j}$ is the geodesic between i and j.

$$L = \frac{1}{n(n-1)} \sum_{i \in N} \sum_{j \in N, j \neq i} d_{ij}. \tag{4.2.2}$$

$$Q = \frac{1}{l} \sum_{ij \in N} \left[a_{ij} - \frac{k_i k_j}{l} \right] \delta(c_i, c_j), \tag{4.2.3}$$

where $l = \sum_{i,j \in N} a_{ij}$, m_i and m_j are the modules containing node i and j, respectively, and $\delta(c_i, c_j) = 1$ if $c_i = c_j$ and 0 otherwise.

$$T = \frac{\sum_{i \in N} 2t_i}{\sum_{i \in N} k_i(k_i - 1)}, \tag{4.2.4}$$

where $t_i = \frac{1}{2} \sum_{j,h \in N} (a_{ij} a_{ih} a_{jh})$ is the number of triangles around node i.

Using the AAL atlas, we constructed the following local brain network metrics at each region or node. We used the local efficiency ($E_{loc,i}$; Equation 4.2.5), clustering coefficient (C_i; Equation 4.2.6) and betweenness centrality (b_i; Equation 4.2.7) at each node to quantify the local connectivity. Both local efficiency and clustering coefficient measure the presence of well-connected clusters around the node, and they are highly correlated to each other. The betweenness centrality is the number of shortest paths which pass through the node, and measures the effect of the node on the overall flow of information in the network (Rubinov and Sporns, 2010). The local connectivity metrics used in this work, for a single node i, are defined as follows;

$$E_{loc,i} = \frac{\sum_{j,h \in N, j \neq i} a_{ij} a_{ih} [d_{jh}(N_i)]^{-1}}{k_i(k_i - 1)}, \tag{4.2.5}$$

where, $d_{jh}(N_i)$, is the length of the shortest path between node j and h - as defined in Equation, and contains only neighbours of h 4.2.1.

$$C_i = \frac{2t_i^w}{k_i(k_i - 1)}.$$

(4.2.6)

$$b_i = \frac{1}{(n-1)(n-2)} \sum_{h,j \in N, h \neq j, h \neq i, i \neq j} \frac{\rho_{hj}(i)}{\rho_{hj}},$$

(4.2.7)

where $\rho_{hj}(i)$ is the weights of shoetest path between h and j that passes throgh i.

4.2.3 Statistical Analysis.

4.2.3 Statistical Analysis. We used different statistical methods as described below, and for the multiple testing we relied on the Bonferroni correction(White et al., 2019; Narum, 2006). Where applicable, the thresholds were obtained by dividing 0.05 by the number of tests.

Quantifying the Change in CDR and Connectivity Metrics

To determine the longitudinal change in CDR, local and global connectivity metrics, we calculated the absolute difference between the first visit (the baseline visit) and the first visit after 12 months (the follow-up visit). Unless stated otherwise, this is the primary way of quantifying this longitudinal change we used in the analysis.

Estimation of Gene Expression from Multiple Probe Sets

Different probe set expression values were present for each gene in the data. To estimate a representative gene expression out of the probe set expression, we conducted a non-parametric Mann-Whitney U test to evaluate whether the expression in AD was different from those of controls. For each gene, we selected the probe set expression that has the lowest Mann-Whitney U p-value. In this way, we selected the most differentially expressed probe sets in our data and considered those for the remaining analysis.

Spearman's Rank Correlation Coefficient

To test the statistical significance of pair-wise undirected relationships, we used the Spearman's rank correlation coefficient (ρ). The Spearman coefficient is a non-parametric method which ranks pairs of measurements and assesses their monotonic relationship. We report here the coefficient ρ along with the corresponding p-value to evaluate the significance of the relationship. A ρ of ± 1 indicates a very strong relationship, while $\rho = 0$ means there is no relationship.

Quantile Regression

To model the directed relationship between two variables, we used the quantile regression model (Koenker and Hallock, 2001). This model is used as an alternative to the linear regression when assumptions of linear regression are not met. This fact allows the response and predictor variables to have non-symmetric distribution. The quantile regression model estimates the conditional median of the dependent variable given the independent variables. Besides, it can be used to estimate any conditional quantile; and is therefore robust to outliers. In this work, we used the second quantile; the median, to model the directed relationship between two variables using the quantile regression.

Ridge Regression

For estimating the relationship between more than two variables, we used ridge regression (Hoerl and Kennard, 1970). The basic idea behind this model is that it solves the least square function penalizing it using the l_2 norm regularization. More specifically, the ridge regression minimizes the following objective function:

$$||y - X\beta||_2^2 + \alpha||\beta||_2^2,$$

$$\text{i.e.,}$$

$$\beta^{Ridge} = \underset{\beta \in \mathbb{R}}{argmin}||y - X\beta||_2^2 + \alpha||\beta||_2^2, \tag{4.2.8}$$

where y is the dependent (or response) variable, X is the independent variable (feature, or predictor), β is the ordinary least square coefficient (or, the slope), α is the regularization parameter, β^{Ridge} is the ridge regression coefficient, $argmin$ is the argument of minimum and it is responsible for making the function attain the minimum and is $L_2(v) = ||v||_2$ represents the L2 norm function (Strang et al., 1993). Moreover, we normalized the predictors to get a more robust estimation of our parameters.

Software

We used Python 3.7.1 for this work; our code has been made available under the MIT License, and is accessible .

4.3 Results

4.3.1 Longitudinal Connectivity Changes and CDR. Initially, we used descriptive statistics plots to visualize the data for the two populations of AD and matched control subjects. To facilitate the integrated analysis, we looked into the different sets of data individually to have a better understanding of the underlying statistical distribution of each, and chose the best analysis methods accordingly. Firstly, we plotted the global and local connectivity metrics in a way that illustrates the longitudinal change. Those longitunal changes are measured after 1 year followup from baseline screening. The global connectivity metric box plots show the baseline and follow-up distributions for both AD and controls for transitivity, Louvain modularity, characteristic path length and global efficiency (Figure 4.2). The figure shows that the longitudinal changes in connectome metrics are statistically significant among the AD subjects and not mere artifacts, but not within the control population which seem to have non significant changes. In fact, comparing all populations values, the only significant differences were for the AD group and for the characteristic path length (p-value 0.0057), global efficiency (p-value 0.0033), and Louvain modularity (p-value 0.0086).

Appendix Figures A1, A2 and A3 show the distribution of the local efficiency, clustering coefficient and betweenness centrality connectivity metrics, respectively, at the baseline and follow-up (left sub-figures), as well as their absolute differences (right sub-figures), at all atlas brain regions. A list of the brain atlas region names, abbreviations and ids are available in Appendix Table A1. Moreover, we show, in Figure 4.3, the scatter and violin plots of the six CDR scores, at the baseline and follow-up. Those are the memory, orientation, judgement and problem solving, community affairs, home and hobbies, and personal care scores which take the categorical values illustrated in the Materials and Methods (and also in Figure 4.3).

Both global and local connectivity features show non-symmetric distribution in the baseline, follow-up and absolute change between them. Therefore, we use non-parametric models and

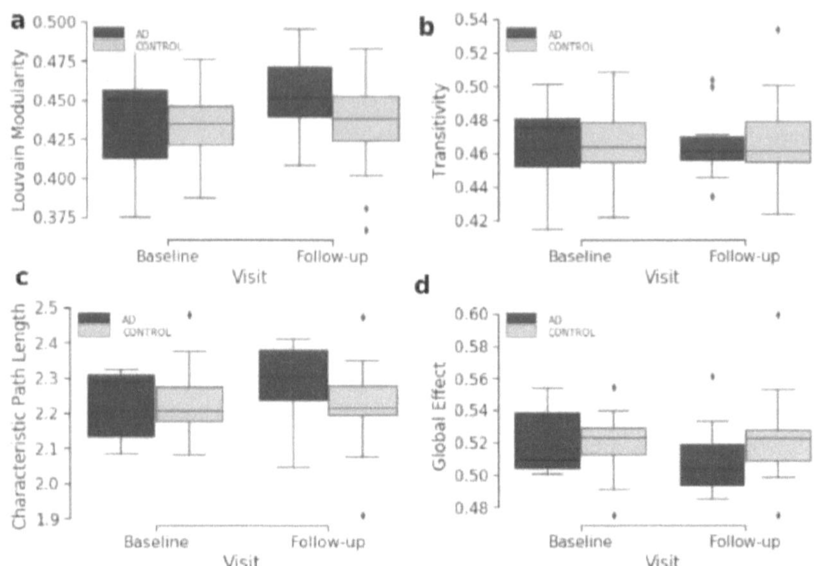

Figure 4.2: Box plots of the distribution of brain segregation and integration global connectivity metrics comparing the two time points. The plots compare the baseline and follow-up distributions for AD and controls for Louvin modularity (a), transitivity (b), characteristic path length (c) and global efficiency (d).

statistical tests in the following analysis.

4.3.2 Gene Expression. We derived a list of 17 AD risk factor genes from BioMart, and retrieved 56 related probes sets. We performed a Mann-Whitney U test which aims at testing whether a specific probe set expression is different between AD and controls. For each gene, we chose the probe set that has the lowest p-value. Table 4.1 reports the selected probe set with the smallest p-value, at each gene. After estimating the expression of the 17 genes, as explained in the Materials and Methods, we plotted a heatmap of the related gene expression profiles showed in Figure 4.4. Here, some of the genes appear to be highly expressed in the profiles (e.g. *SORL1* and *PSEN1*), while others show very low expression (e.g. *HFE* and *ACE*).

4.3.3 Association Analysis. We studied the undirected associations of the 17 gene expression with the longitudinal change in global and local brain connectivity, as well as the associations with longitudinal CDR and connectivity changes. The total sample size after integrating all the datasets was 47 participants. Firstly, we performed an association analysis of gene expression with the connectivity changes locally, at each Automated Anatomical Labeling (AAL) brain region. In

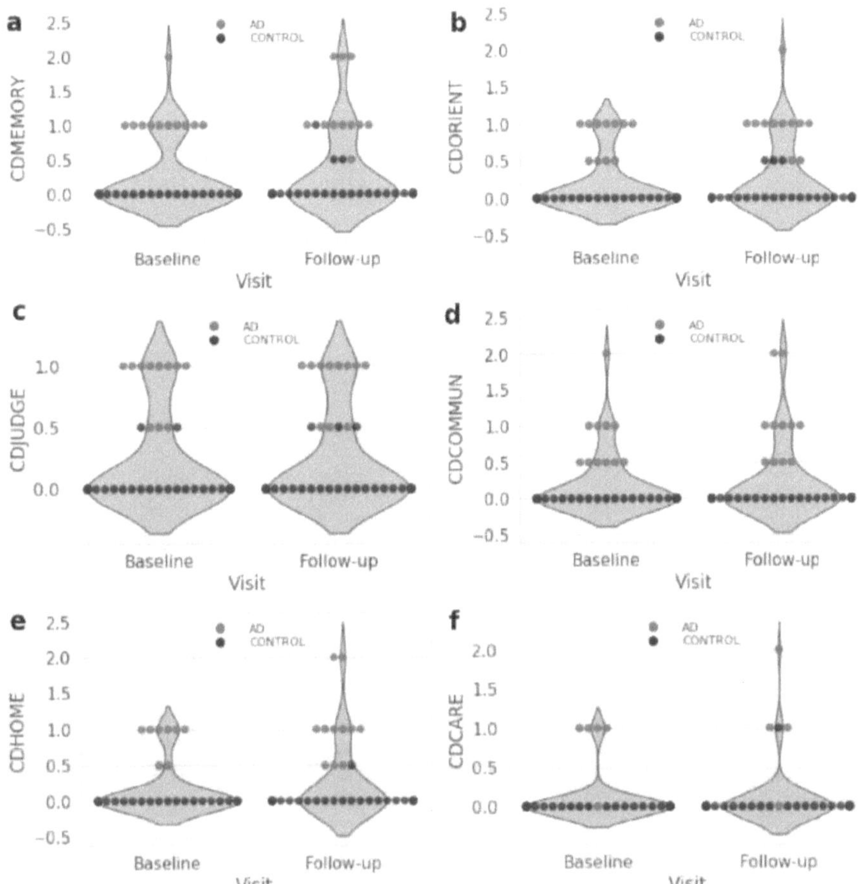

Figure 4.3: Violin plots to illustrate the CDR scores (either 0: None, 0.5: very mild, 1: mild, 2: moderate or 3: severe) in the baseline (left violin plot) and follow-up (right violin plot) visits, for AD (red dots) and controls (blue dots). The memory (CDMEMORY; a) and orientation (CDORIENT; b) scores are represented by the top sub-figures, judgment and problem solving (CDJUDGE; c) and community affairs (CDCOMMUN; d) are the middle sub-figures, while home and hobbies (CDHOME; e) and personal care (CDCARE; f) are at the bottom. It is visible that generally some AD subjects worsen their score, except for the CDCARE score where few improved as a result to finding strategies after the diagnosis at baseline.

Table 4.2 we show the top results reported along with the Spearman correlation co-efficient. The *APP* gene, ρ =-0.58, p-value=1.9e-05) and *BLMH*, ρ =0.57, p-value=2.8e-05) are the top and

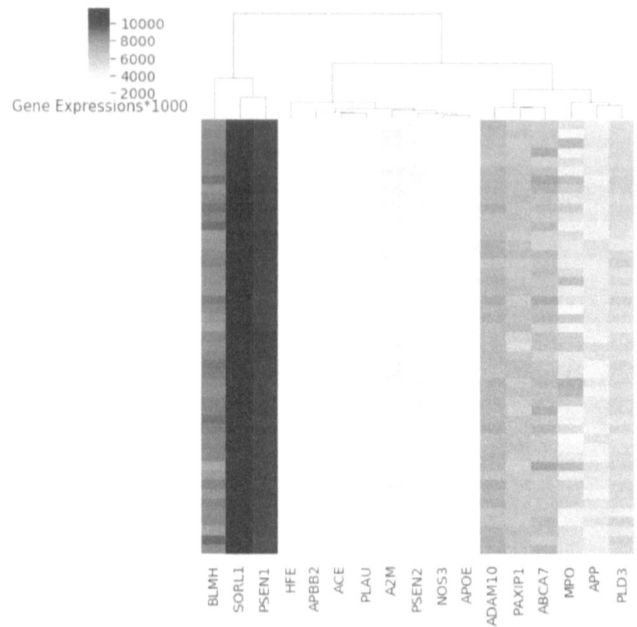

Figure 4.4: A heatmap of the estimated 17 gene expressions (values multiplied by 1000, each line represents a participant) out of the 65 probe sets as explained in the Materials and Methods section. The dark blue represents a high expression values, while the yellow represents low expression. The *SORL1* has the highest expression among the genes and *HFE* expression was the lowest among other genes.

only significant genes in the list, and associate with the change in local efficiency at the right middle temporal gyrus (Temporal_Mid_R AAL region) and clustering coefficient at the left Heschl gyrus (Heschl_L), respectively. Figure 4.5 shows the scatter plots related to the latter scenarios.

In Table 4.2, there is a similar pattern observed in association results between the clustering coefficient and local efficiency, e.g. both metrics are associated with *BLMH* at the left Heschl gyrus (Heschl_L), *APP* at the right middle temporal gyrus (Temporal_Mid_R) and *PLAU* at the right ngular gyrus (Angular_R). We interpret this by the strong correlation that exists between the local efficiency and clustering coefficient, at the baseline, follow-up and also, the absolute change (see Appendix Figure A4). On the other hand, Appendix Table A2 reports the top results of the association between gene expression and the change in brain global connectivity. In this case there are no significant associations.

Table 4.1: Mann-Whitney U test top results for the difference between AD and controls in probe-set expression

Gene	Top results		
	Chromosome	Probe set id	p-value
APBB2	4	11734823_a_at	0.02575
MPO	17	11727442_at	0.38631
APP	21	11762804_x_at	0.01396
ACE	17	11752871_a_at	0.24478
PLAU	10	11717154_a_at	0.01396
PAXIP1	7	11755176_a_at	0.45499
HFE	6	11736346_a_at	0.11881
SORL1	11	11743129_at	0.10912
A2M	12	11715363_a_at	0.28592
NOS3	7	11725467_a_at	0.04261
BLMH	17	11757556_s_at	0.09356
ADAM10	15	11751180_a_at	0.14278
PLD3	19	11715382_x_at	0.17304
ApoE	19	11744068_x_at	0.05962
PSEN1	14	11718678_a_at	0.29453
PSEN2	1	11723674_x_at	0.04862
ABCA7	19	11755091_a_at	0.45499

4.3.4 Regressing Change in Local and Global Brain Connectivity on Gene Expression.

We analyzed the directed association through regressing the change in local connectivity (as a dependant variable), at each AAL region, on gene expression using (as an independent variable or predictor) a quantile regression model. Table 4.3 reports the top results, along with the regression coefficient, p-values and t-test statistic. *PLAU* was the most significant gene affecting the absolute change in betweenness centrality at left Fusiform gyrus (Fusiform_L) with an increase of 487.13 at each unit increase in *PLAU* expression (p-value= $3e - 06$). This was followed by the expression of *HFE* with an effect size of 0.1277 on the change in local efficiency at the right anterior cingulate and paracingulate gyri (Cingulum_Ant_R). Those observed associations are illustrated in Figure 4.6. Appendix Figures A5, A7 and A6 show the Manhattan plots for the -log10 of the p-values corresponding to the quantile regression models of the change in local efficiency, clustering coefficient and betweenness centrality, respectively.

Similarly, we regressed the absolute change of global connectivity measures on gene expression values and the top results are shown in Appendix Table A3. All the results have p-values less than the threshold we set ($\frac{0.05}{17} = 0.0029$).

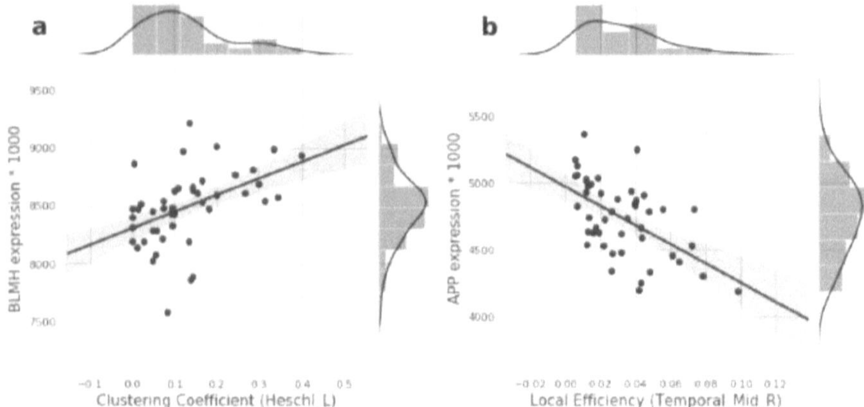

Figure 4.5: A scatter plot of all the significant association results. The plots shows the associations between; (a) *BLMH* expression and clustering coefficient in AAL region 79 (Heschl_L), (b) *APP* expression and local efficiency in brain region 86 (Temporal_Mid_R).

Table 4.2: Top results of Spearman associations between AD gene expression and local connectivity metrics.

Gene	Sorted by P-value. Dashed line: threshold$=\frac{0.05}{17\times90}=3.27e-05$				
	Region	Region id	Metric	ρ	P-value
APP	Temporal_Mid_R	Region86	local_eff	-0.5805	1.9e-05
BLMH	Heschl_L	Region79	cluster_coef	0.5708	2.8e-05
PSEN1	Occipital_Mid_R	Region52	b_centrality	-0.5598	4.3e-05
BLMH	Heschl_L	Region79	local_eff	0.5591	4.4e-05
APP	Temporal_Mid_R	Region86	cluster_coef	-0.5197	0.000182
PAXIP1	Amygdala_L	Region41	cluster_coef	0.5064	0.000281
PLAU	Angular_R	Region66	cluster_coef	0.484	0.000567
PLAU	Angular_R	Region66	local_eff	0.4838	0.00057
ACE	Postcentral_L	Region57	b_centrality	0.4648	0.000998
ADAM10	Postcentral_L	Region57	local_eff	-0.4602	0.001133
PAXIP1	Parietal_Sup_L	Region59	b_centrality	-0.4585	0.00119
PLAU	Fusiform_L	Region55	b_centrality	0.4564	0.001262
SORL1	Putamen_R	Region74	local_eff	-0.4528	0.001395
PSEN2	Frontal_Inf_Oper_R	Region12	local_eff	0.4457	0.001693
PLAU	Frontal_Inf_Oper_L	Region11	b_centrality	0.4454	0.001704
ABCA7	Temporal_Inf_L	Region89	local_eff	0.442	0.001866

Table 4.3: Top 50 quantile regression results of the change in local network metrics (y) on and targeted Alzheimer's Disease gene expression (x)

Region	Sorted by p-value. Dashed line: threshold$=\frac{0.05}{17\times90}=3.27e-05$					
	Region	R. id	Beta	Statistic	P-value	Metric
PLAU	Fusiform_L	55	487.1319	5.3836	3e-06	b_centrality
HFE	Cingulum_Ant_R	32	0.1277	4.8139	1.7e-05	local_eff
PAXIP1	Parietal_Sup_L	59	-147.3175	-4.5608	3.9e-05	b_centrality
HFE	Cingulum_Ant_R	32	0.1662	3.9835	0.000246	cluster_coef
APP	Amygdala_R	42	-0.1349	-3.8548	0.000365	local_eff
PLAU	Hippocampus_L	37	0.1073	3.4801	0.001125	local_eff
ADAM10	Postcentral_L	r57	-0.0871	-3.4376	0.001275	cluster_coef
ApoE	Frontal_Inf_Orb_L	15	153.3117	3.3627	0.001584	b_centrality
APBB2	Amygdala_L	41	0.2054	3.3517	0.001635	cluster_coef
ApoE	Frontal_Sup_Medial_L	23	0.1912	3.2788	0.002015	cluster_coef
MPO	Cingulum_Mid_L	33	0.0293	3.2465	0.00221	cluster_coef
MPO	Cingulum_Mid_L	33	0.0281	3.2143	0.00242	local_eff
PLAU	Cingulum_Ant_R	32	0.1428	3.1969	0.002543	local_eff
ADAM10	Postcentral_L	57	-0.0517	-3.1541	0.002867	local_eff
ApoE	Postcentral_L	57	0.0806	3.0931	0.003398	local_eff
PLD3	Olfactory_R	22	-0.1268	-3.0463	0.003867	cluster_coef
ABCA7	Frontal_Inf_Orb_R	16	34.9538	2.9489	0.005043	b_centrality
A2M	Putamen_R	74	-0.0543	-2.9171	0.005495	local_eff
PLAU	Hippocampus_L	37	0.1472	2.9023	0.005717	cluster_coef
HFE	Frontal_Inf_Tri_R	14	0.1852	2.8813	0.006047	cluster_coef
HFE	Frontal_Inf_Tri_R	14	0.0926	2.8594	0.006411	local_eff
APP	Amygdala_R	42	-0.1753	-2.8288	0.006953	cluster_coef
ApoE	Occipital_Mid_R	52	44.0624	2.7995	0.007512	b_centrality
HFE	Calcarine_R	44	0.1403	2.7916	0.007669	cluster_coef
APP	Temporal_Mid_R	86	-0.0692	-2.7396	0.008787	cluster_coef
APP	Temporal_Mid_R	86	-0.0386	-2.7297	0.009016	local_eff
PLD3	Olfactory_R	22	-0.0987	-2.713	0.009413	local_eff
A2M	Olfactory_R	22	-30.9342	-2.6919	0.009941	b_centrality
APP	Cuneus_R	46	0.1443	2.6845	0.010131	cluster_coef
PSEN1	Frontal_Inf_Tri_L	13	-0.1344	-2.6492	0.01109	local_eff
PSEN2	Temporal_Mid_L	85	0.0528	2.6465	0.011168	local_eff
PSEN1	Frontal_Inf_Tri_L	13	-0.2431	-2.6432	0.011262	cluster_coef
PLAU	Frontal_Mid_R	8	0.0854	2.6384	0.011401	local_eff
ApoE	Putamen_L	73	372.8291	2.638	0.011411	b_centrality
A2M	Occipital_Mid_R	52	0.0535	2.6213	0.011907	cluster_coef
APP	Cuneus_R	46	0.0798	2.6189	0.011979	local_eff
ADAM10	Temporal_Sup_L	81	-0.1123	-2.6134	0.012147	cluster_coef
ApoE	SupraMarginal_L	63	0.1158	2.5663	0.013677	local_eff
HFE	Calcarine_R	44	0.0755	2.533	0.014866	local_eff
PLAU	Occipital_Mid_L	51	0.1268	2.5193	0.01538	cluster_coef
MPO	Pallidum_R	76	0.024	2.5101	0.015732	local_eff
ABCA7	Temporal_Inf_L	89	0.0249	2.5083	0.015804	local_eff
A2M	Occipital_Mid_R	52	0.0268	2.5023	0.01604	local_eff
PLAU	Hippocampus_R	38	182.0756	2.5007	0.016105	b_centrality
APP	Frontal_Med_Orb_L	25	0.1167	2.4968	0.01626	cluster_coef
HFE	Cingulum_Post_R	36	28.3207	2.4708	0.017334	b_centrality
ACE	Occipital_Mid_R	52	0.0433	2.4632	0.017659	local_eff
NOS3	Olfactory_L	21	-0.1615	-2.4596	0.017815	cluster_coef
ABCA7	Temporal_Inf_L	89	0.0429	2.4374	0.018808	cluster_coef
SORL1	Paracentral_Lobule_L	69	91.7443	2.4364	0.018855	b_centrality

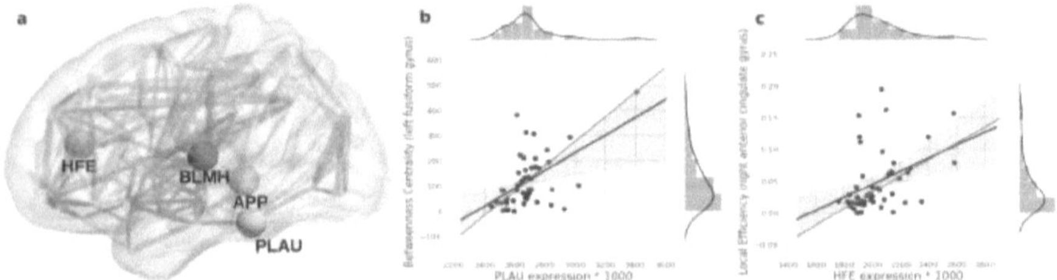

Figure 4.6: Subfigure (a) higlights regions in the brain where significant associations - between gene expression and longitudinal change in local connectivity metrics - were found, using quantile regression (*HFE* and *PLAU*) and spearman associations (*APP* and *BLMH*). Each gene is plotted at the AAL brain region where the association was significant; *APP* at Temporal_Mid_R, *BLMH* at Heschl_L, *PLAU* at Fusiform_L and *HFE* at Cingulum_Ant_R. (b) and (c) are scatter plots to visualize the association between *PLAU* gene expression and betweenness centrality in the left fusiform gyrus (a), and between the expression of *HFE* gene with local efficiency in right anterior cingulate gyrus (b). The red line on the plots represents the median (quantile) regression line, while the blue line represents the ordinary least square line.

4.3.5 Additive Genetic Effect on Brain Regions. To visualize the overall contribution of AD gene risk factors used in this work on distinct brain areas, we added up the -log10 p-values for the gene expression coefficients at each of the 90 AAL regions. The p-values were obtained from the quantile regression analysis between the gene expression values and each of the three connectivity metrics - those are the absolute difference between baseline and follow-up of local efficiency, clustering coefficient and betweenness. Figure 4.7 summarizes this by 1) representing the brain connectome without edges for each one of the connectivity metric, 2) each node represents a distinct AAL region and is annotated with the name of the region, 3) the size of each node is the sum -log10 of the regression coefficient associated p-vales for all the genes. The color is assigned automatically by the BrainNet Viewer. Overall, although the gene contributions to the absolute change in local efficiency have a similar pattern to that of clustering coefficient, the contribution to betweenness centrality change is relatively small.

4.3.6 Regressing the difference in CDR on the difference in Global and Local Connectivity. To asses the directed and undirected association of the longitudinal measures of global connectivity and CDR scores, we calculated the difference between baseline and follow-up visits for both CDR and global connectivity metrics , i.e. $CDR_{baseline} - CDR_{follow-up}$ and $metric_{baseline} - metric_{follow-up}$, respectively. The Spearman and quantile regression results are both shown in Table 4.4. We observe that the increase in overall brain segregation - through transitivity- reduces the memory over time ($\beta = -6.14e - 06$, p-value= 0.0034). On the other hand, there is a positive association between the brain integration - through global efficiency-

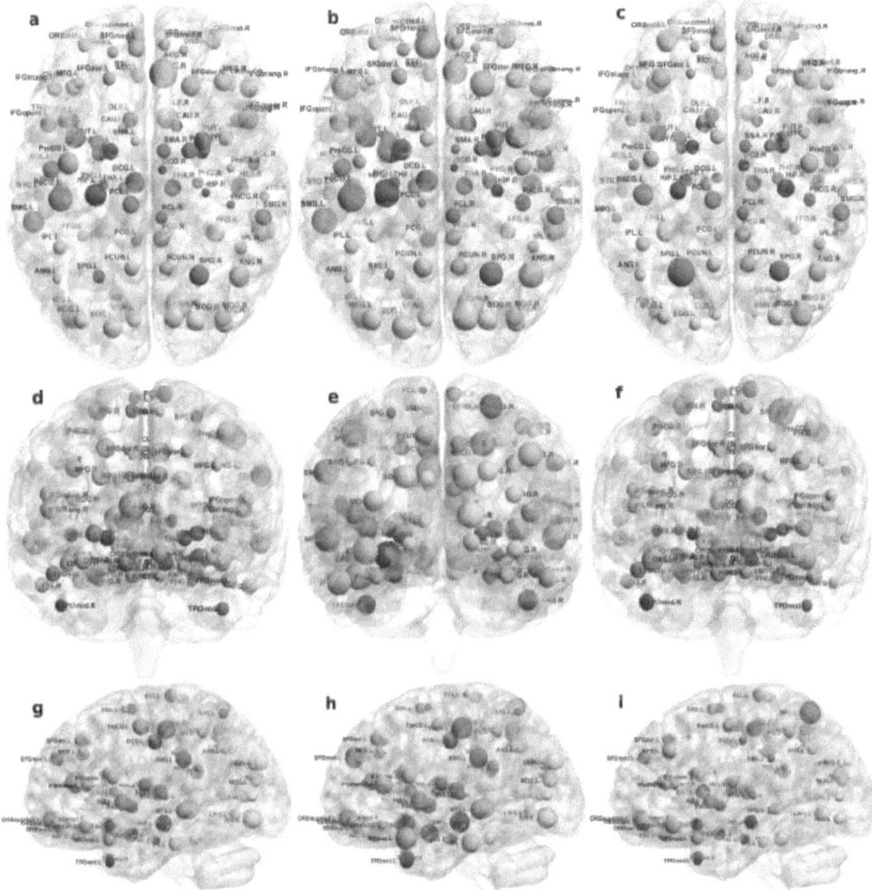

Figure 4.7: Connectome representations showing the metric additive genetic effect at each AAL node. The subfigures show the axial (top; (a), (b) and (c)), coronal (middle; (d), (e) and (f)), and sagittal (bottom; (g), (h) and (i)) planes of the brain, the node size represents the local efficiency (left; (a), (d) and (g)), clustering coefficient (middle; (b), (e) and (h)) and betweenness centrality (right; (c), (f) and (i)). Colors of the nodes are automatically assigned by the BrainNet Viewer. The acronyms of the brain regions are explained in Appendix Table A1.

and home and hobbies.

Similarly, in Appendix Table A4 we looked at the monotonic effect of local connectivity metrics on

Table 4.4: Quantile regression results of the difference in CDR (y) with the difference in global connectivity (x)

Global Metric	Threshold= $\frac{0.05}{6} = 0.00833$. * represents significant p-value.					
	CDMEMORY	CDORIENT	CDJUDGE	CDCOMMUN	CDHOME	CDCARE
Q. Regression: β **(p-value)**						
transitivity	-6.14e-06 (0.0034*)	-1.8e-06 (nan)	-3.2e-07 (nan)	8.4e-07 (0.9249)	4.13e-06 (0.7324)	-3.1e-07 (nan)
global_eff	1.3e-06 (0.8572)	3.36e-06 (0.9944)	3.5e-07 (0.9826)	3.09e-06 (0.3131)	9.5e-06 (0.1613)	-2.5e-07 (nan)
louvain	-2.64e-06 (0.1683)	-6.84e-06 (0.0352)	1.21e-06 (0.0012*)	-1.01e-06 (0.8125)	-2.16e-06 (0.9424)	-6.3e-07 (0.1787)
char_path_len	-5.8e-07 (0.8562)	-8.3e-07 (0.8791)	-8e-08 (0.8637)	-8.3e-07 (0.0361)	-2.14e-06 (0.0442)	-2e-08 (nan)
Spearman: ρ **(p-value)**						
transitivity	-0.3395 (0.021)	-0.0763 (0.6142)	-0.0618 (0.6835)	0.0661 (0.6623)	0.161 (0.2851)	-0.0081 (0.9574)
global_eff	0.0483 (0.7497)	0.0056 (0.9707)	0.0505 (0.7388)	0.2685 (0.0712)	0.4145 (0.0042*)	-0.0263 (0.8625)
louvain	-0.1955 (0.1928)	-0.3119 (0.0349)	0.2077 (0.166)	-0.0968 (0.522)	-0.11 (0.4666)	-0.0628 (0.6784)
char_path_len	-0.1183 (0.4337)	-0.0516 (0.7333)	-0.0281 (0.8531)	-0.2909 (0.0498)	-0.3811 (0.009)	-0.0065 (0.9658)

the seven CDR scores, both represented as the subtraction of the follow-up visit from the baseline visit. The increase in betweenness centrality was shown to have different effects on the CDR score over the one-year time period. For example, as the betweenness centrality decreases over time, the judgement and problem solving increases in severity by 1.06e-08 over time (p-value=1.32e-17), in the frontal lobe (Frontal_Inf_Oper_L).

4.3.7 Multivariate Analysis: Ridge Regression. Additionally, we regressed the difference in CDR visits (response variable; Y), one score at a time, on both the difference in global brain connectivity (predictor; X1), one connectivity metric at a time and all gene expression values (predictor; X2), using the ridge regression model. Table 4.5 reports the mean squared error (the score column) and shows the top hits in the multiple ridge regression. It shows that the α (alpha column) could not converge, using the cross-validation, when the response variables were the judgment or personal care. However, the CDR score results show that genes and connectivity metrics have a small effect (β) on the response variables (the change in CDR scores over time), and the larger effects were observed when using the total CDR score (CDR_diff) as a response variable. The expression of genes have negative and positive effects on CDR change, and so are the connectivity metrics. The expression of *ApoE*, for example, has a negative effect (β) of -0.24 on the change in memory score, i.e. the memory rating decreases by 0.24 as the *ApoE* expression increases. While if the expression of *ApoE* increases one unit, the home and hobbies score increases, over time, by 0.12.

Table 4.5: Ridge regression results of the change CDR scores on the global connectivity changes and Alzheimer's Disease gene expressions.

CDR	Metric	Alpha	score	Metric	APBB2	MPO	APP	ACE	PLAU	PAXIP1	HFE	SORL1	A2M	NOS3	BLMH	ADAM10	PLD3	ApoE	PSEN1	PSEN2	ABCA7
CDMEMORY	transitivity	0.1	-0.0747	-0.6704	0.122	0.0823	0.0248	-0.0392	-0.0556	0.0497	-0.0926	0.0138	-0.3465	0.3276	-0.1321	-0.1347	-0.0774	-0.2426	0.161	-0.1369	0.1903
CDMEMORY	global_eff	0.1	-0.0753	-0.0304	0.1114	0.0802	0.0332	-0.0311	-0.0575	0.06	-0.0765	0.0086	-0.3486	0.3226	-0.1364	-0.1318	-0.0628	-0.267	0.1585	-0.1392	0.1919
CDMEMORY	louvain	0.1	-0.0754	-0.259	0.1052	0.0823	0.0353	-0.0259	-0.0676	0.0647	-0.0634	0.0098	-0.3483	0.3178	-0.1388	-0.1296	-0.0616	-0.2694	0.1563	-0.1373	0.1884
CDMEMORY	char_path_len	0.1	-0.0752	-0.1203	0.1004	0.0804	0.0351	-0.0335	-0.0617	0.0619	-0.0769	0.0074	-0.3507	0.3187	-0.1365	-0.1287	-0.0475	-0.2827	0.1621	-0.1398	0.1869
CDORIENT	transitivity	0.1	-0.0677	-0.3061	0.1731	0.0679	0.2135	0.0287	0.0046	-0.0399	-0.472	0.0156	-0.192	0.0553	-0.0138	-0.0342	-0.0135	-0.1967	0.1309	-0.232	-0.0035
CDORIENT	global_eff	0.1	-0.0677	-0.0517	0.169	0.067	0.2172	0.0324	0.0042	-0.0354	-0.4648	0.0133	-0.1928	0.0532	-0.0157	-0.0332	-0.0078	-0.2068	0.1297	-0.2331	-0.0025
CDORIENT	louvain	0.1	-0.0673	-0.403	0.1592	0.0702	0.2203	0.0406	-0.0117	-0.0281	-0.4445	0.0151	-0.1923	0.0457	-0.0195	-0.0298	-0.0058	-0.2107	0.1261	-0.23	-0.0079
CDORIENT	char_path_len	0.1	-0.0677	0.009	0.1688	0.067	0.2173	0.0325	0.0039	-0.0352	-0.4645	0.0132	-0.1929	0.0532	-0.0158	-0.033	-0.0076	-0.2071	0.1295	-0.2331	-0.0025
CDJUDGE	transitivity	149	-0.0056	-0.0	-0.0002	0.0023	-0.0007	-0.0012	0.0	0.0019	0.0003	0.0002	0.0004	0.0002	-0.0005	-0.0002	-0.0007	-0.0003	-0.0005	-0.0001	-0.0027
CDJUDGE	global_eff	149	-0.0056	0.0	-0.0002	0.0023	-0.0007	-0.0012	0.0	0.0019	0.0003	0.0002	0.0004	0.0002	-0.0005	-0.0002	-0.0007	-0.0003	-0.0005	-0.0001	-0.0027
CDJUDGE	louvain	149	-0.0056	0.0001	-0.0002	0.0023	-0.0007	-0.0012	0.0	0.0019	0.0003	0.0002	0.0004	0.0002	-0.0005	-0.0002	-0.0007	-0.0003	-0.0005	-0.0001	-0.0027
CDJUDGE	char_path_len	149	-0.0056	-0.0001	-0.0002	0.0023	-0.0007	-0.0012	0.0	0.0019	0.0003	0.0002	0.0004	0.0002	-0.0005	-0.0002	-0.0007	-0.0003	-0.0005	-0.0001	-0.0027
CDCOMMUN	transitivity	0.3	-0.0325	-0.0526	0.1227	-0.0205	-0.178	0.1151	0.1812	-0.0161	-0.0366	-0.0375	-0.1628	0.1476	-0.0795	0.1688	-0.0516	0.0563	0.0078	-0.1135	0.0637
CDCOMMUN	global_eff	0.3	-0.0325	0.0071	0.1218	-0.0207	-0.1775	0.1154	0.1811	-0.0155	-0.0356	-0.0378	-0.1629	0.1473	-0.0798	0.1692	-0.0503	0.0547	0.0076	-0.1137	0.0637
CDCOMMUN	louvain	0.3	-0.0325	0.0174	0.1223	-0.0208	-0.1776	0.1152	0.1817	-0.0158	-0.0363	-0.0379	-0.1629	0.1476	-0.0797	0.169	-0.0505	0.0548	0.0077	-0.1138	0.064
CDCOMMUN	char_path_len	0.5	-0.0324	-0.0703	0.1038	-0.0164	-0.1436	0.0941	0.1423	-0.0117	-0.0184	-0.0381	-0.1579	0.1242	-0.0666	0.1484	-0.0458	0.0447	0.0139	-0.1111	0.0613
CDHOME	transitivity	0.4	-0.0893	0.0428	0.1908	-0.0343	-0.09	0.0756	0.0009	-0.0142	0.0578	-0.1228	-0.1029	-0.0044	-0.1787	0.3326	-0.095	0.1188	0.113	-0.3091	0.1076
CDHOME	global_eff	0.3	-0.0891	0.0872	0.1961	-0.0363	-0.0987	0.0821	-0.0017	-0.0156	0.0521	-0.132	-0.1063	-0.0057	-0.1846	0.3515	-0.0929	0.1258	0.1227	-0.3232	0.1079
CDHOME	louvain	0.3	-0.0892	-0.0191	0.1972	-0.0361	-0.0989	0.0826	-0.0017	-0.0157	0.0525	-0.1318	-0.1059	-0.0054	-0.1847	0.3511	-0.0949	0.1275	0.1225	-0.3231	0.1082
CDHOME	char_path_len	0.6	-0.0891	-0.1843	0.1674	-0.0302	-0.0767	0.0607	0.0025	-0.0114	0.0616	-0.1083	-0.0978	-0.0083	-0.167	0.3047	-0.0782	0.0932	0.1009	-0.2832	0.0988
CDCARE	transitivity	149	-0.1093	-0.0	-0.0008	0.0027	-0.0014	-0.0002	0.0007	0.0001	-0.0037	0.0013	-0.0066	-0.001	-0.0037	-0.0018	-0.0031	0.0	0.0013	-0.0045	0.0078
CDCARE	global_eff	149	-0.1093	-0.0	-0.0008	0.0027	-0.0014	-0.0002	0.0007	0.0001	-0.0037	0.0013	-0.0066	-0.001	-0.0037	-0.0018	-0.0031	0.0	0.0013	-0.0045	0.0078
CDCARE	louvain	149	-0.1093	-0.0001	-0.0008	0.0027	-0.0014	-0.0002	0.0007	0.0001	-0.0037	0.0013	-0.0066	-0.001	-0.0037	-0.0018	-0.0031	0.0	0.0013	-0.0045	0.0078
CDCARE	char_path_len	149	-0.1093	0.0	-0.0008	0.0027	-0.0014	-0.0002	0.0007	0.0001	-0.0037	0.0013	-0.0066	-0.001	-0.0037	-0.0018	-0.0031	0.0	0.0013	-0.0045	0.0078
CDGLOBAL	transitivity	0.1	-0.0477	-0.3476	0.1049	0.0302	-0.1995	0.0836	-0.0396	-0.0238	0.206	0.168	-0.0559	0.213	-0.1703	-0.0285	-0.0784	0.0136	-0.0127	-0.1342	0.0448
CDGLOBAL	global_eff	0.1	-0.0479	0.1446	0.0964	0.0292	-0.1943	0.0875	-0.0422	-0.0176	0.2152	0.1649	-0.0579	0.2093	-0.1728	-0.0259	-0.0666	-0.0032	-0.0136	-0.1356	0.0446
CDGLOBAL	louvain	0.1	-0.0479	-0.1177	0.0965	0.0301	-0.1942	0.0902	-0.0452	-0.0163	0.2203	0.1658	-0.0569	0.2082	-0.1736	-0.026	-0.0702	-0.0002	-0.015	-0.1345	0.044
CDGLOBAL	char_path_len	0.3	-0.0479	-0.2835	0.0738	0.0313	-0.1599	0.059	-0.0278	-0.0008	0.1636	0.1121	-0.0542	0.1526	-0.1467	-0.0237	-0.0363	-0.0125	0.0167	-0.1282	0.0396
CDR_diff	transitivity	0.1	-1.2761	-1.5541	0.7401	0.1734	-0.2195	0.1158	0.197	0.0516	-0.8229	-0.0517	-1.2177	0.8435	-0.6461	0.3132	-0.3519	-0.337	0.6769	-1.3473	0.6318
CDR_diff	global_eff	0.1	-1.2782	0.1652	0.7112	0.1687	-0.1988	0.134	0.1901	0.0768	-0.7842	-0.0645	-1.2239	0.8302	-0.6564	0.3216	-0.3117	-0.3999	0.6719	-1.353	0.634
CDR_diff	louvain	0.1	-1.28	-0.4664	0.7041	0.1724	-0.1962	0.1439	0.1741	0.084	-0.762	-0.0618	-1.2221	0.823	-0.6605	0.3239	-0.3153	-0.3984	0.6672	-1.3492	0.6292
CDR_diff	char_path_len	0.1	-1.2776	-1.0277	0.6252	0.1706	-0.1849	0.1144	0.1581	0.0904	-0.7903	-0.0733	-1.2397	0.8002	-0.6563	0.3454	-0.1921	-0.5225	0.7013	-1.3571	0.5943

4.4   Discussion

Our results show that Alzheimer's risk genes can manipulate the amount of change observed in the structural connectome, measured as the absolute difference of longitudinal connectivity metrics. Here, we show that longitudinal regional connectivity metrics, global brain segregation and integration have effects on the CDR scores. More specifically, we observe a consistent decrease, over time, in the local efficiency - a connectivity metric that measures the efficient flow of information around a node (a brain region) in its absence (Rubinov and Sporns, 2010) - in response to the increase in *APP* expression, at the right middle temporal gyrus (Temporal_Mid_R; see Table 4.2). The same connectivity metric increases over time as the expression of *HFE* increases, at the right anterior cingulate and paracingulate gyri (see Table 4.3). Furthermore, as the disease progresses, we observe a correlation between brain segregation and cognitive decline, the latter is measured as CDR memory scores. While if the brain becomes more integrated, as measured by global efficiency; it results in an improved growth of home and hobbies scores (see Table 4.4).

Prescott et al. (2014) have investigated the differences in the structural connectome in three clinical stages of AD, using a cross-sectional study design, and targeted regional brain areas that are known to have increased amyloid plaque. Their work suggested that connectome damage might occur at an earlier preclinical stage towards developing AD. Here, we further adapted a longitudinal study design and incorporated known AD risk genes. We showed how the damage in the connectome is associated with gene expression, and that the change in connectome affects dementia, globally and locally - at distinct brain regions. Aside from our previous work (Elsheikh et al., 2018a), which examined the ApoE associations with longitudinal global connectivity in AD using longitudinal global connectivity metrics, this study, to the best of our knowledge, is the first of its type to include gene expression data with global and local brain connectivity. However, similar work has been done in schizophrenia structural brain connectivity, where longitudinal magnetic resonance imaging features, derived from the DTI, were associated with higher genetic risk for schizophrenia (Alloza et al., 2018).

The results obtained here align with findings in the literature of genetics and neuroimaging. Specifically, Robson et al. (2004) studied the interaction of the *C282Y* allele *HFE* - the common basis of hemochromatosis - and found that carriers of *ApoE-4*, the C2 variant in *TF* and *C282Y* are at higher risk of developing AD. Moreover, the *HFE* gene is known for regulating iron absorption, which results in recessive genetic disorders, such as hereditary haemochromatosis (Pilling et al., 2019). According to Pujol et al. (2002), the association between the harm avoidance trait and right anterior cingulate gyrus volume was statistically significant. In their study, they examined the association between the morphology of cingulate gyrus and personality in 100 healthy participants. Personality was assessed using the Temperament and Character Inventory questionnaire. Higher levels of harm avoidance were shown to increase the risk of developing AD (Wilson et al., 2011). We show here that *HFE* expression affects the local efficiency at the right anterior cingulate gyrus (see Table 4.3 and Figure 4.6). This might indicate a possible effect of *HFE* expression on the personality of AD patient or the person at risk of developing the disease.

Moreover, in this study we found that the Plasminogen activator, urokinase (*PLAU*) expression affects the betweenness centrality (a measure of the region's (or node) contribution to the flow of information in a network (Rubinov and Sporns, 2010)) in the left fusiform gyrus, over time (see Table 4.3 and Figure 4.6). Although the functionality of this region is not fully understood, its relationship with cognition and semantic memory was previously reported (Galton et al., 2001). *PLAU*, on the other hand, was shown to be a risk factor in the development of late-onset AD as a result of its involvement in the conversion of plasminogen to plasmin - a contributor to the processing of *APP* - by the urokinase-type plasminogen activator (*uPA*) (Finckh et al., 2003).

When examining the linear associations between gene expression and local connectivity (see Table 4.2 and Figure 4.5), we found that the right middle temporal gyrus, known for its involvement in cognitive processes including comprehension of language, negatively associates with *APP* expression. Additionally, the left Heschl gyrus positively correlates with bleomycin hydrolase (*BLMH*) expression. In the human brain, the *BLMH* protein is found in the neocortical neurons and senile plaques (Namba et al., 1999), microscopic decaying nerve terminals around the amyloid occurring in the brain of AD patients. Some studies (Papassotiropoulos et al., 2000; Farrer et al., 1998) have found that a variant in the *BLMH* gene, which leads to the Ile443→Val in the *BLMH* protein, increases the risk of AD; this was strongly marked in *ApoE-4* carriers. The BLMH protein can process the $A\beta$ protein precursor and is involved in the production of $A\beta$ peptide (Kajiya et al., 2006).

Even though none of the AD risk genes showed a significant effect on the longitudinal change in global connectivity (see Appendix Tables A3 and A2), the genes showed significant effects on local connectivity changes at regional brain areas (see Table 4.3 and Table 4.2). The global connectivity metrics of the brain, on the other hand, have shown promising results in affecting the change observed in CDR scores, including memory, judgement and problem solving, as well as home and hobbies, as shown in Table 4.4. Previous work studied the association between genome-wide variants and global connectivity of Alzheimer's brains, represented as brain integration and segregation, and found that some genes affect the amount of change observed in global connectivity (Elsheikh et al., 2020b). This suggests that a generalisation of the current study at a gene-wide level might warrant further analysis.

Our work provides new possible insights, though replication on a larger sample size is required. Indeed, one limitation here was the small sample size available. We needed to narrow down our selection of participants to those attended both baseline and follow-up visits, and have CDR scores, genetic and imaging information available. Another limitation is given by the use of only two time points, the baseline and the first follow-up visit. This does not allow capturing the effects of connectivity changes in a longer-term or studying the survival probabilities in AD. Extending to more time points would have been useful, but it would have further reduced the dataset. We foresee future work in using a more complex unified multi-scale model, to facilitate studying the multivariate effect of clinical and genetic factors on the connectome, besides considering the complex interplay of genetic factors.

In summary, in this chapter we conducted an association analysis of targeted gene expression with various longitudinal brain connectivity features in AD. Aiming at estimating the neurodegeneration of the connectome, we obtained local and global connectivity metrics at two visits, baseline and follow-up, after 12 months. We calculated the change between the two visits and carried out an association analysis, using quantile and ridge regression models to study the relationship between gene expression and disease progression globally and regionally at distinct areas of the brain. We tested the effect of the change in connectivity on the longitudinal CDR scores through quantile regression. Furthermore, using a ridge regression model, we controlled for the genetic effects in the previous settings to study the effect of connectivity changes on the CDR change.

The present analysis was conducted in AD using a longitudinal study design and highlighted the role of *PLAU, HFE, APP* and *BLMH* in affecting how the pattern information is propagated in particular regions of the brain, which might have a direct effect on a person's recognition and cognitive abilities. Furthermore, the results illustrated the effect of brain structural connections on memory and cognitive process of reaching a decision or drawing conclusions. The findings presented here might have implications for better understanding and diagnosis of the cognitive deficits in AD and dementia.

Chapter 5

BiGen: Integrative Clinical and Brain-Imaging Genetics Analysis Using Structural Equation Model

5.1   Introduction

Imaging genetics is a rapidly growing field that focuses on the identification of genetic variants associated with complex brain diseases. In doing so, imaging genetics combines brain imaging technology output with genetic data. Depending on the hypothesis of interest and data availability, imaging genetics studies can focus on 1) the association between each location in the brain, namely voxel, with each single nucleotide polymorphism (SNP), e.g. Voxelwise Genome-Wide Association Study (vGWAS) (Stein et al., 2010; Shen et al., 2010), 2) the relationship between a single phenotype, such as the hippocampus volume, these studies are called candidate phenotype studies (Stein et al., 2012), or 3) consideration of multiple imaging endophenotype associations with candidate genotypes.

Recently, brain connectivity and other magnetic resonance imaging (MRI) features have been used as phenotypes in GWAS and other omic studies. Brain connectivity metrics are derived from the connectome - a representation of the brain as a single network extracted from the diffusion tensor imaging (DTI) (Sporns et al., 2005). The nodes of a connectome represent distinct regions in the brain and the links are the water tracts connecting each pair of regions. Some studies used GWAS with connectivity metrics to study the healthy brain (JahanshAD et al., 2013), Alzheimer's Disease (AD) (Elsheikh et al., 2020b), and other brain diseases (Alloza et al., 2018; Thompson et al., 2010). Imaging genetics uses a variety of study designs that integrate specific genetic information (Thompson et al., 2010), in this book and paper Elsheikh et al. (2019a), for example, studied the effect of targeted gene expression on local and global connectivity features in AD.

The recent imaging genetics literature has proposed multivariate methods to understand the influence of multidimensional genotypes on multi-phenotype imaging characteristics. As summarised by Liu and Calhoun (2014); the multivariate methods in the field of imaging genetics need to consider three factors; 1) the dimensionality of genotypes and intermediate neuroimaging phenotypes, 2) the importance of managing the confounding factors in order to reveal the imaging phenotypic effects, 3) the population structure of the sample under study. Considering these criteria, previous studies proposed multivariate analysis methods utilising sparse reduced-rank regression (Vounou et al., 2010), sparse partial least square (Le Floch et al., 2012), parallel independent and principal component analysis (Liu et al., 2009) and sparse canonical correlation analysis (Chi et al., 2013). Other methods considered longitudinal multivariate imaging genetics pipelines to study the relationship of gene expression and genome-wide variants with brain connectivity metrics (Elsheikh et al., 2018a, 2020b, 2019a; Lu et al., 2017).

The structural equation model (SEM) (Bollen and Long, 1993) is a multivariate technique that studies the complex causal relationship between a set of endogenous and exogenous latent constructs through measured or observed variables. SEM estimates the structural relationship between the latent variables using a set of multiple regression models and factor analysis. Recently, SEM has been applied in the context of mapping genotype and phenotype in complex diseases and imaging genetics. Specifically, SEM is a promising tool in merging gene regulatory networks with post-GWAS approaches to improve the understanding of cell signaling, and metabolic pathway translation into phenotypes (Nuzhdin et al., 2012). Grotzinger et al. (2018) proposed Genomic SEM, a tool that identifies between-diseases similarity and genetic architecture, with an application to psychiatric disorders. Huisman et al. (2018) applied the SEM to understand the spatial change within brain regions affected by gene expression using a dataset from Alzheimer's Disease Neuroimaging Initiative (ADNI). They used healthy brain information from spatial transcriptome datasets to identify the model construct.

Here, we propose a method to identify the causal effect of genetics on multiple phenotypes, including neuroimaging and disease measures. We utilised the structural equation model to estimate the association between various latent constructs, including disease, endophenotypes, covariants and genetics. In our model, the latent variables were estimated using observed measurements and the relationship between disease and genetics was estimated accounting for the intermediate effect of neuroimaging endophenotype. The latent genetic variable was measured through observed gene expression, considering the interaction of proteins, while the disease and endophenotype latent variables were measured through the Clinical Dementia Ratings (CDR) and global connectivity metrics, respectively. Moreover, we controlled for confounding factors through a separate latent construct, inferred from some environmental factors. We applied the new tool to simulated data and a dataset from ADNI.

5.2 Materials and Methods

In this section we describe our proposed SEM model, BiGen. The model has two parts, the structural model and the measurement model which are described in the following sub-sections. Thereafter we discuss the estimation of these models.

5.2.1 Structural Model. The structural model in a SEM relates the exogenous and endogenous latent variables (construct) and studies their associations. The advantage of using latent variables is to control for the measurement errors on the overall SEM. In BiGen, the structural model consists of four latent variables, these are genetics, confounders, endophenotypes and disease. These four latent constructs are measured through some genetic measurements, environmental

and other risk factors, neuroimaging characteristics and disease measures of interest. See Figure
5.1 where latent variables are shown as circles.

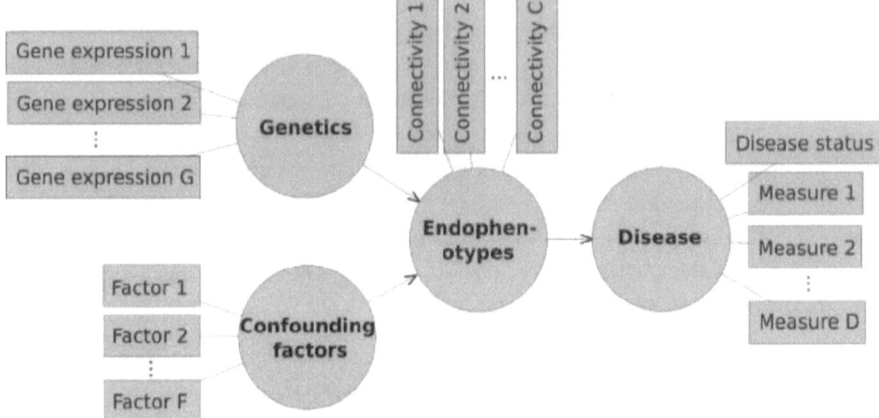

Figure 5.1: The BiGen model showing the latent variables (in circles) and the observed measurements (rectangles). The structural model shows latent variables and the relationship between
them, while the measurement models connects the measurement variables with latent constructs.

5.2.2 Measurement Model. The measurement model focuses on the relationship between
latent variables and their observed measurements (indicators). The measurements are shown as
rectangles in Figure 5.1. Specifically, in BiGen, the genetics and confounding factors constructs
relate to the endophenotypes. The endophenotypes have one direct relationship with the disease
construct. Moreover, the endophenotypes mediate between genetic (and confounding factors)
and the disease construct. This mediation is considered to account for the indirect effects of
genetic factors on the disease through the endophenotypes. BiGen analyses the effect of genetic-
endophenotype interaction on the disease.

5.2.3 Model Estimation. We applied the BiGen model to an AD dataset from ADNI. In
this case, the constructs were measured and inferred from a set of observed measurements.
Specifically, the genetics construct was measured through gene expression values that interact
with one another. These gene expression values correlate the measurements of other latent
variables with the genetics construct. The confounding factors are the random effects or other risk
factors that are not genetics. Here, other risk factors, if available, could be used as measurements
for the latter construct, such as blood pressure, cholesterol level and heart diseases. A set of
global connectivity metrics were used to measure the endophenotypes construct. In this work
we used the global connectivity metrics, namely, global efficiency, Louvain modularity, transitivity

and characteristics path length. Finally, the disease construct was measured through a set of clinical dementia ratings. Section 5.2.4 and Section 5.2.5 explain the simulated and AD datasets in more detail, respectively.

The structural model consists of two exogenous constructs (Figure 5.1), these are genetics and confounding factors, and two endogenous constructs, endophenotypes and disease. To estimate the SEM we followed two estimation steps. In the first step, we estimated the latent variable scores through an iterative algorithm that does not assume any distribution of the measurements or constructs. In doing so, we adjusted the partial least squares (PLS) SEM iterative algorithm proposed by Lohmöller (1989). The approach estimates the latent scores in four iterative steps until a stop criteria is met. The second step estimates the path coefficients and outer weights through ordinary least square models. This basic PLS algorithm has been applied successfully to many fields, including advertising research (Henseler et al., 2012). In our model, we used non-parametric methods and a summary of the steps we adopted is provided below.

In our first step, we followed a set of sub-steps that aim to estimate the latent scores iteratively until convergence. More specifically, considering the notation shown in Figure 5.2, the first sub-step is to estimate $w_i j$ for all the measurement models connecting $x_i j$ with y_j. Initially, we initialised these weights to 1, and updated them iteratively. The second sub-step is to estimate the inner weights (b_{ik}; see Figure 5.2). This is followed by an approximation of the latent variables (y_i^*), and in the last sub-step we estimate the outer weights. These four sub-steps are summarised below. The tolerance is calculated at the end of each iteration, and this is simply the difference between the current outer weight and the previous one. The threshold here was set to $1e - 6$.

1. External approximation of latent variables, $y_i = \sum wx$

2. Estimation of inner weights, $b_{ik} = \rho_{ik}$, then normalize.

3. Internal approximation of latent variables, $y_i^* = \sum by$

4. Estimation of outer weights, $w = \rho_{xy^*}$ (where ρ is Spearman correlation coefficient), then normalize.

As a second step, BiGen estimates the final path coefficients and outer weights through a (quantile) regression model. Besides the inner model shown in Figure 5.1, BiGen calculates the inner of the interaction terms: genetics×endophenotype and confounding factors×endophenotype and estimates their effect on the disease.

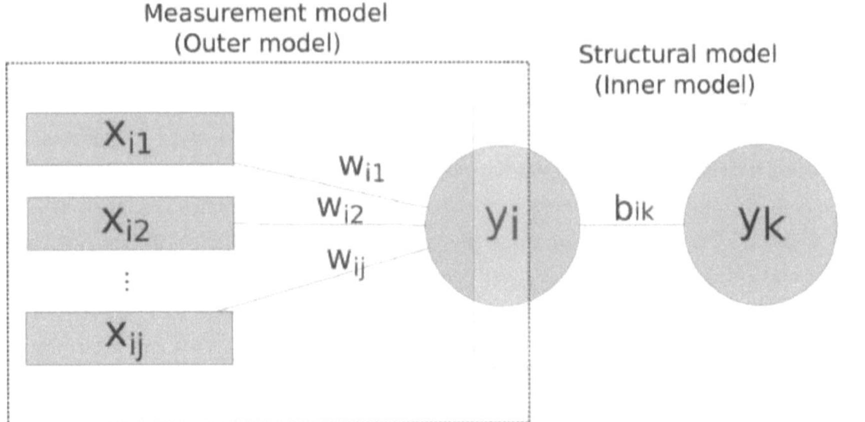

Figure 5.2: Notation used in model estimation. The left box represents the measurement model, and the right box represents the structural model. The outer weights $(w_{i1}, w_{i2} \ldots w_{ij})$ connect the indicators $(x_{i1}, x_{i2} \ldots x_{ij})$ to the latent variable y_i. The inner weight (b_{ik}) associates the latent variables in the structural model (y_i and y_k).

5.2.4 Simulated Data. We simulated a dataset to test BiGen and compare it to the PLS-SEM. Firstly, we randomly generated a single continuous measurement variable of the disease construct. We then created another two measurement variables with strong correlation with the first measure, but with some randomness. We simulated three connectivity variables that are associated with one another, and to the three measures of the disease measurements, with some randomness. Similarly, we simulated the genetic and confounding factor measures (three for each).

To simulate the variables, we used a random range of parameters. Specifically, we simulated 12 variables as described above, each follow a random distribution with different mean (ranges between 100 and 300) and standard deviation (ranges between 2 and 18). Figure 5.3 shows the distribution of the simulated data, while the pattern of association between the variables is shown in Figure 5.4. The sample size of the simulated data is 5000.

5.2.5 Alzheimer's Disease Dataset. We also applied our model to an AD dataset from ADNI . Specifically, we used the information of ADNI participants as measurement variables to fit our proposed model (as shown in Figure 5.1). For the 1) disease, 2) endophenotypes, 3) confounding factors and 4) genetics we used the following measurements; 1) three CDR measures, namely, memory, judgment and problem solving, and home and hobbies scores, 2) the global connectivity metrics, namely, Louvain modularity, transitivity, global efficiency

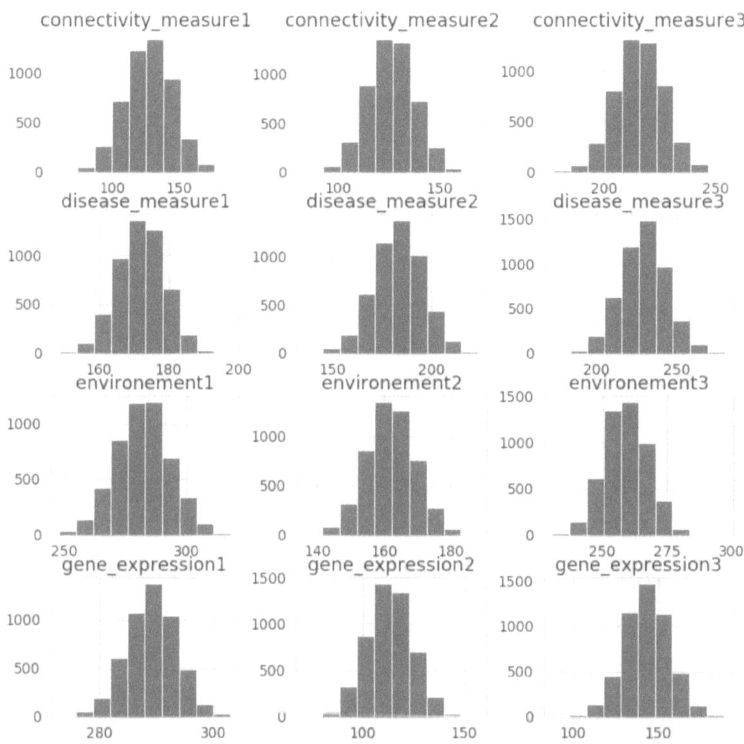

Figure 5.3: The distribution of all the simulated measurement variables used to fit the proposed BiGen model.

and characteristics path length (all explained in Section 4.2.2), 3) the education level and gender of participants, and 4) the expression of a set of genes that interact with one another, namely *APP, SORL1, ADAM10, ApoE* and *PSEN1*. We obtained the expression values following the same steps explained in 4.3.2. The protein-protein interaction data were derived from STRING database (v11.0), accessed at (Szklarczyk et al., 2014, 2018). We used the absolute difference between the baseline and follow-up visits for the disease and endophenotypes measurements. Figure 5.5 shows the distribution of all pre-described measurements, while Figure

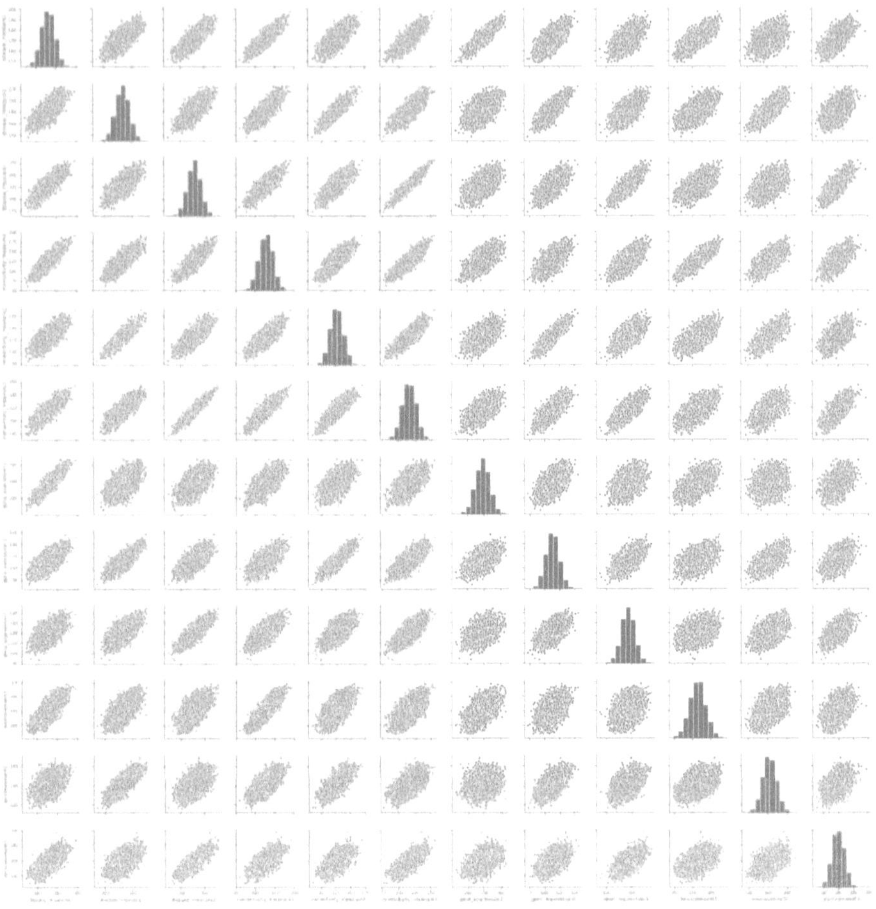

Figure 5.4: The association pattern between all the measurement variables from the simulated data.

5.6 shows their association patterns. In total, we managed to obtain 46 samples.

Software

To conduct the analysis described here, we used Python 3.7.1 and made our scripts available under the MIT License and accessible .

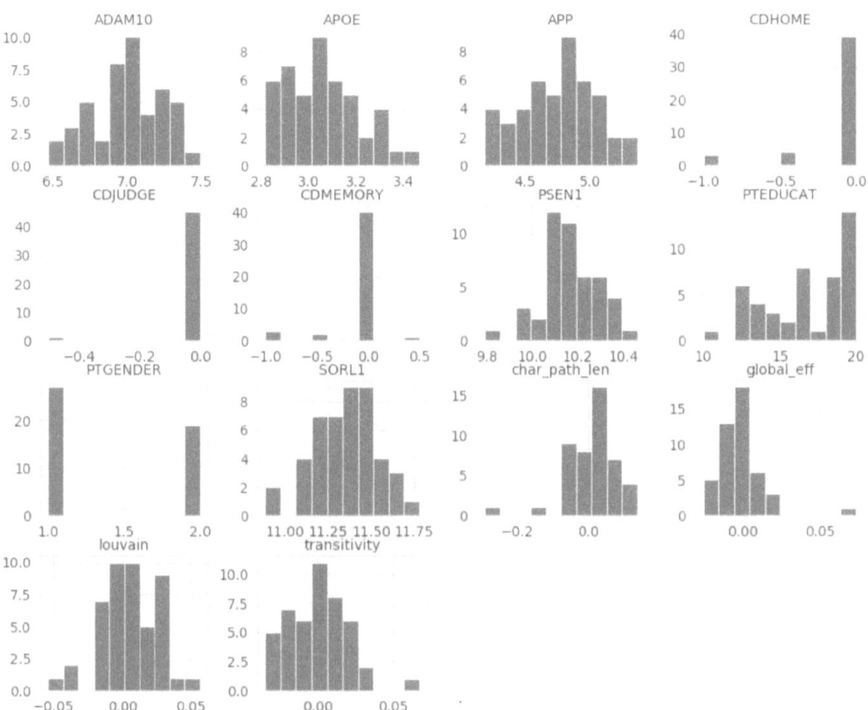

Figure 5.5: The distribution of all the measurement variables obtained from ADNI dataset and used to fit the proposed BiGen model.

5.3   Results

5.3.1 Evaluation using Simulated Data.

PLS-SEM

Using the simulated measurement variables described in Section 5.2.4 we fitted, firstly, the PLS-SEM, the path coefficient and inner weights are both shown in matrices 5.3.1, and the interaction terms (described in Section 5.2.3) inner weights are shown in matrix 5.3.2. We then fitted BiGen (as described in 5.2.3) to the same data and obtained the path and inner weights that are shown in matrices 5.3.3, similarly, we obtained the inner weights estimates for the interactions, this is shown in matrix 5.3.4.

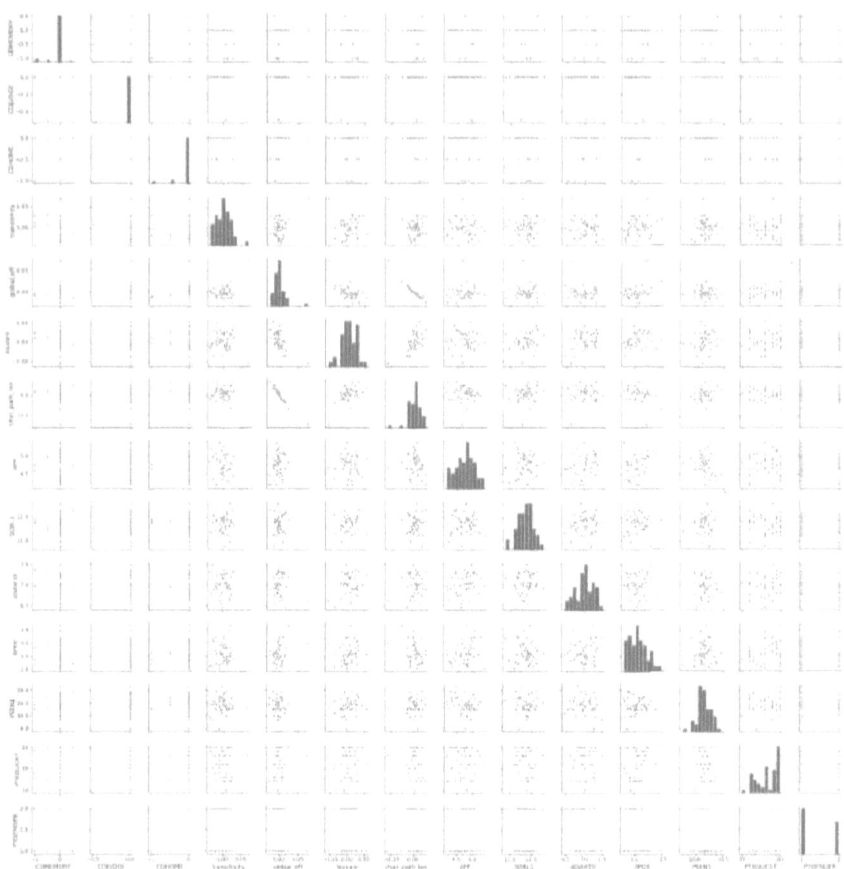

Figure 5.6: The association patterns between all the measurement variables obtained from ADNI dataset.

$$
\text{outer} = \begin{pmatrix}
Disease1: & 0.353 & 0. & 0. & 0. \\
Disease2: & 0.371 & 0. & 0. & 0. \\
Disease3: & 0.378 & 0. & 0. & 0. \\
Endo1: & 0. & 0.359 & 0. & 0. \\
Endo2: & 0. & 0.347 & 0. & 0. \\
Endo3: & 0. & 0.357 & 0. & 0. \\
Gene1: & 0. & 0. & 0.360 & 0. \\
Gene2: & 0. & 0. & 0.429 & 0. \\
Gene3: & 0. & 0. & 0.414 & 0. \\
Enviro1: & 0. & 0. & 0. & 0.432 \\
Enviro2: & 0. & 0. & 0. & 0.411 \\
Enviro3: & 0. & 0. & 0. & 0.399
\end{pmatrix}
\quad
\text{inner} = \begin{pmatrix}
Disease: & 0. & 0. & 0. & 0. \\
Endo: & 0.959 & 0. & 0. & 0. \\
Genetics: & 0. & 0.878 & 0. & 0. \\
Enviro & 0. & 0.892 & 0. & 0.
\end{pmatrix}
$$

$$(5.3.1)$$

PLS-SEM wtih Interaction

$$\text{inner} \begin{pmatrix} Disease: & 0. & 0. & 0. \\ Gene\_Endo: & -0.0293 & 0. & 0. \\ Enviro\_Endo: & -0.0153 & 0. & 0. \end{pmatrix} \qquad (5.3.2)$$

BiGen

$$\text{outer} = \begin{pmatrix} 0.353 & 0. & 0. & 0. \\ 0.372 & 0. & 0. & 0. \\ 0.378 & 0. & 0. & 0. \\ 0. & 0.360 & 0. & 0. \\ 0. & 0.345 & 0. & 0. \\ 0. & 0.357 & 0. & 0. \\ 0. & 0. & 0.358 & 0. \\ 0. & 0. & 0.430 & 0. \\ 0. & 0. & 0.414 & 0. \\ 0. & 0. & 0. & 0.432 \\ 0. & 0. & 0. & 0.412 \\ 0. & 0. & 0. & 0.397 \end{pmatrix} \quad \text{inner} = \begin{pmatrix} 0. & 0. & 0. & 0. \\ 0.958 & 0. & 0. & 0. \\ 0. & 0.882 & 0. & 0. \\ 0. & 0.897 & 0. & 0. \end{pmatrix} \qquad (5.3.3)$$

BiGen wtih Interaction

$$\text{inner} = \begin{pmatrix} 0. & 0. & 0. \\ 0.01230 & 0. & 0. \\ -0.00260 & 0. & 0. \end{pmatrix} \qquad (5.3.4)$$

5.3.2 Application to ADNI. Then, we applied both BiGen and SEM-PLS to the AD measurement variables discussed in Section 5.2.5. Similarly, we considered the interaction model in both applications. The path coefficient and inner weights obtained from applying the PLS-SEM are both shown in matrices 5.3.5, while the interaction inner weights are shown in matrix 5.3.6. We then fitted BiGen to the AD dataset and the results of the path and inner weights are shown in matrices 5.3.7. We finally computed the inner weights estimates for the interactions (see matrix 5.3.8).

Generally, we observe that applying BiGen and PLS-SEM to the simulated data gave the same results (compare: matrices 5.3.1 vs matrices 5.3.3, and matrix 5.3.2 vs matrix 5.3.4), however, in the AD application the two models gave slightly, though not significantly different results (compare: matrices 5.3.5 vs matrices 5.3.7, and matrix 5.3.6 vs matrix 5.3.8). We observe that BiGen produced different effects of the inner path (inner matrix 5.3.7) compared to PLS-SEM (inner matrix 5.3.5) when applied to ADNI dataset. The overall effect of the genetic and confounding factor latent constructs on the endophenotype are 0.113 and 0.216 in BiGen, while they are -0.337 and -0.150 in PLS-SEM. Though none of these were statistically significant, we observe that the contribution of genetic factors to the endophenotype are lower in BiGen, the positive sign indicates the direction of the relationship. This might indicate that PLS-SEM overestimates the inner path effect of genetic construct on the endophenotype and underestimates the confounding factor effect on the endophenotype.

PLS-SEM

$$
\text{outer} = \begin{pmatrix}
Memory: & 2.958\text{e-}01 & 0. & 0. & 0. \\
Judge: & -6.640\text{e-}04 & 0. & 0. & 0. \\
Home: & 9.039\text{e-}01 & 0. & 0. & 0. \\
Transitivity: & 0. & -9.517\text{e-}02 & 0. & 0. \\
Global\_eff: & 0. & -4.127\text{e-}01 & 0. & 0. \\
Louvain: & 0. & 1.640\text{e-}01 & 0. & 0. \\
Char\_path\_len: & 0. & 4.520\text{e-}01 & 0. & 0. \\
APP: & 0. & 0. & -1.659\text{e-}01 & 0. \\
SORL1: & 0. & 0. & 1.720\text{e-}01 & 0. \\
ADAM10: & 0. & 0. & 5.762\text{e-}01 & 0. \\
ApoE: & 0. & 0. & -7.060\text{e-}01 & 0. \\
PSEN1: & 0. & 0. & 3.970\text{e-}01 & 0. \\
Educat: & 0. & 0. & 0. & 9.896\text{e-}01 \\
Gender: & 0. & 0. & 0. & -2.137\text{e-}02
\end{pmatrix}
\tag{5.3.5}
$$

$$
\text{inner} = \begin{pmatrix}
Disease: & 0. & 0. & 0. & 0. \\
Endo: & 0.280 & 0. & 0. & 0. \\
Genetics: & 0. & -0.337 & 0. & 0. \\
Enviro: & 0. & -0.150 & 0. & 0.
\end{pmatrix}
$$

PLS-SEM wtih Interaction

$$
\text{inner} = \begin{pmatrix}
Disease: & 0. & 0. & 0. \\
Gene\_Endo: & -0.1212 & 0. & 0. \\
Enviro\_Endo: & 0.1658 & 0. & 0.
\end{pmatrix}
\tag{5.3.6}
$$

BiGen

$$
\text{outer} =
\begin{pmatrix}
Memory: & 0.390 & 0. & 0. & 0. \\
Judge: & -0.211 & 0. & 0. & 0. \\
Home: & 0.839 & 0. & 0. & 0. \\
Transitivity: & 0. & 0.063 & 0. & 0. \\
Global\_eff: & 0. & -0.380 & 0. & 0. \\
Louvain: & 0. & 0.310 & 0. & 0. \\
Char\_path\_len: & 0. & 0.426 & 0. & 0. \\
APP: & 0. & 0. & -0.554 & 0. \\
SORL1: & 0. & 0. & 0.292 & 0. \\
ADAM10: & 0. & 0. & 0.079 & 0. \\
ApoE: & 0. & 0. & -0.386 & 0. \\
PSEN1: & 0. & 0. & 0.444 & 0. \\
Educat: & 0. & 0. & 0. & 1.009 \\
Gender: & 0. & 0. & 0. & 0.948
\end{pmatrix}
\quad
\text{inner} =
\begin{pmatrix}
Disease: & 0. & 0. & 0. & 0. \\
Endo: & 0.207 & 0. & 0. & 0. \\
Genetics: & 0. & 0.113 & 0. & 0. \\
Enviro: & 0. & 0.216 & 0. & 0.
\end{pmatrix}
\tag{5.3.7}
$$

Similarly, the interaction terms were smaller when we fit BiGen to the ADNI dataset compared to PLS-SEM (compare the BiGen inner interaction terms in matrix 5.3.8 vs SEM-PLS inner interaction terms in matrix 5.3.6). Our overall explanation of these results, including the simulation results, is that the simulated data are made to be symmetric and normally distributed, therefore, both the PLS-SEM and BiGen models offer a similar performance, while in the ADNI application, the pattern of distribution is not symmetric and the data types are different, which is always the case in real life applications. However, we believe that more samples are needed to better understand the performance of BiGen in the ADNI dataset.

Table 5.1 shows a brief comparison of the PLS-SEM and BiGen results in the application to simulation and ADNI datasets. We note that both BiGen and PLS-SEM were fast (see the Time column), even though the ADNI application needed more iterations in both models. To get the interaction terms, the same calculations in the original model are performed, except that we need to estimate different path coefficients in the second step (see Section 5.2.3).

BiGen with Interaction

$$
\text{inner}
\begin{pmatrix}
Disease: 0. & 0. & 0. \\
Gene\_Endo: 0.0014 & 0. & 0. \\
Enviro\_Endo: -0.0068 & 0. & 0.
\end{pmatrix}
\tag{5.3.8}
$$

Table 5.1: SEM, BiGen with and without interaction results for simulated and AD dataset.

Application	Simulation			AD		
	Iterations	Tolerance	Time[*,‡]	Iterations	Tolerance	Time
SEM	2	$7.918e-08$	202 ms[+]	23	$6.702e-07$	241 ms
SEM (interaction)	2	$7.918e-08$	189 ms	23	$6.702e-07$	184 ms
BiGen	2	$1.367e-07$	502 ms	26	0.000	1.32 s[†]
BiGen (Interaction)	2	$1.367e-07$	521 s	26	0.000	1.26 s

[*] Running time
[+] Millisecond
[†] Second
[‡] Program was ran using Jupyter nootebook in a Debian 9 operating system, model: Intel(R) Core(TM) i5-3230M CPU @ 2.60GHz.

The main difference between the above results and what was done in Chapters 3 and 4 is that here we did not explicitly test the pairwise associations of the observed measurement variables (e.g, global connectivity vs gene expression), but rather, we tested their combined association through the latent constructs (e.g. the effect of genetic on the endophenotype), this was finally estimated by the inner weights, shown in the matrices above. For example, Table A2 shows that global connectivity metrics associated with the expression of *APP*, *ADAM10*, *ApoE* and *PSEN1* with correlation coefficients -0.2602 (characteristic path length), -0.236 (Louvain modularity), -0.25 (characteristic path length) and -0.299 (transitivity). On the other hand, BiGen can not measure the association in the same way, but rather, 1) we observe the contribution of the previous gene expression on the genetic construct, these are $-0.554, 0.079, -0.386, 0.444$, respectively (see outer matrix 5.3.7), 2) BiGen also allows measuring the effect of the genetic construct on the endophenotype, and that was 0.113 (see inner matrix 5.3.7).

In Chapter 3 we conducted a GWAS on the ADNI dataset and integrated the summary statistics obtained from GWAS at a gene-wide level, we found that some genes were significantly associated with global connectivity metrics (e.g *ANTXR2*, *IGF1* and *OR5L1*). However, to study the effect of these genes using BiGen, we considered the protein-protein interactions, and incorporate other factors shown in Figure 5.1. Additionally, In Chapter 4, we tried to test different hypotheses individually, we tested the effect of gene expression on the connectivity metric using both quantile regression and Spearman association coefficient, we then examined the effect of brain connectivity on the CDR measurements, after which we tested their combined effect on the CDR measurements using ridge regression. BiGen allows us to test all these hypotheses (including the interaction terms between them) simultaneously in only two steps and with less computational complexity.

5.4   Conclusion

The SEM is a commonly multivariate technique used in many scientific fields, it studies the structural relationships between variables that are hard to measure directly. Analysing imaging genetics data needs not only a multivariate technique, but also one that considers the complexity, heterogeneity and multicollinearity of both the imaging and genetics parts. Here we propose BiGen, a Python tool that analyses the structural relationship between four constructs simultaneously, these are, *genetics* and *disease*, mediated by the *neuroimaging endophenotypes* and considering some *environmental or confounding factors*. All the four latent constructs are measured through observed indicators and the model is specifically made for brain-related disease, with an application to Alzheimer's disease through ADNI data. BiGen adjusts the PLS-SEM by using non-parametric regression and association tests in estimating the path model and latent variable scores, respectively. This causal predictive model is flexible, simple and computationally inexpensive. Additionally, BiGen has a satisfactory performance in small samples and assumes no distribution or hypothesis of the data.

Our model can easily be extended to include more indicators and constructs. For example, one can add more imaging phenotypes, or rather, incorporate the local connectivity metrics to the endophenotype latent construct. Moreover, the model is flexible to more complexity in the path model, this is especially useful when studying the direct relationship between genetics and disease phenotypes, or the gene-environment interaction effects on the disease or endphenotypes. Over all, we propose that BiGen is suitable for unveiling the complex interplay between neuroimaging and genetics in the context of brain-related diseases.

Although there are some limitations of SEM related to causality inference as a result of fitting multiple linear models which require certain assumptions to be met (Rockman, 2008), our model tries to avoid some through normalisation of the variables. However, the linearity assumptions of the relationship between the dependent and independent variable remains unsolved here.

Chapter 6

General Discussion

In this book, we studied a brain disease with local lesions (we considered GBM), then we looked at a widespread brain disease which affects the overall brain structure, we considered AD in this part. We tested different hypotheses to understand the association between neuroimaging characteristics (e.g. brain connectivity and tumour texture and spatial characteristics) and multi-omic factors (including genome-wide variations and transcriptomic data). We observed the need for one unified model to study the complex interplay between genetic, environmental and clinical, neuroimaging and phenotype features. We introduced a novel model which can test complex hypotheses in the field of imaging genetics, study the effect of interaction terms on the final phenotype, and accommodate heterogeneous data types including phenotype measurements, brain connectivity, environmental and multi-omic factors. Our model was fast, simple, flexible and assumes no distribution of the data.

Our brains are vulnerable to different disorders that affect our everyday functionality, mental and cognition abilities. Neurodegenerative diseases attack the neurones and consequently reflect in a fast cognitive decay. Alzheimer's disease (AD) is a neurodegnarative disease with an average survival of 4 - 8 years after diagnosis (Gaugler et al., 2019). In the early disease stages, biological changes in the brain start occurring slowly, these include the abnormal accumulation of tau beta-amyloid proteins. AD progression is relatively slow at first, however, it worsens over time as it approaches the later stages. The cognitive decline starts to become obvious and the change in brain structure becomes measurable through MRI technologies. Even though AD's risk increases with age, AD is rather a complex disease which develops as a result of many factors, including genetics.

The ongoing advancement in brain imaging techniques coupled with the vast genetic data available presents an opportunity for advancing research in the field of imaging genetics. Genome-wide

Association Studies (GWAS) have shown success in identifying the genetics of Alzheimer's disease (Lambert et al., 2013; Escott-Price et al., 2014). Moreover, recent discoveries in neuroscience have provided the field with better characterization of the brain through connectomics Hagmann et al. (2008). The connectome summarises the brain as one network, extracted from the diffusion tensor/weighted imaging. The nodes of a connectome segment the brain into distinct regions that are connected through edges, these edges are formed/weighted by the water tracts connecting each pair of regions. In many ways, connectomics are used to quantify and study the structural connectivity in the brain; with a special success in understanding the mechanisms and characteristics of AD (Brown et al., 2011; Elsheikh et al., 2018a). Using connectomic phenotypes in GWAS and next generation sequencing research has successfully revealed disease-associated variants in AD Cuyvers and Sleegers (2016). Despite the progress made, we lack an understanding of how the AD connectome progresses over time in relation to the genetic factors.

While neurodegenerative diseases affect connectivity across the brain, brain tumors, on the other hand, are a mass of abnormal cells, which originate in the brain and affect particular regions. Malignant brain tumors are fast growing, and often recur after resection. Glioblastoma multiforme (GBM) is the fourth grade of glioma brain tumor, which is malignant. It is caused by the glial cells that support the nerve cell functionality in the brain. GBM is hard to treat and patients have a median survival time of 11-15 months, eventually, it leads to death. In this project, we also studied the diagnosis and prognosis in a highly heterogeneous population of patients with glioblastoma (Elsheikh et al., 2018b). We used several radiomic methods to identify genetic differences between tumors, by means of non-parametric Spearman associations. We used a publicly available dataset from the Cancer Genome Atlas (TCGA) and the associated brain imaging from The Cancer Imaging Archive (TCIA). Both datasets are well-vetted and publicly available. We used the multi-modal MRI scans, provided by TCIA, to compute the spatial and texture features of GBM tumor, and correlate this with gene-wide expression data. Our results found a group of genetic markers, including *EPGN*, *LRRC46*, *TCN1*, *OR2AE1*, *TUBA1C* and *ZNF284*, whose expression strongly correlate with some of spatial and radiomic tumor features, though no significant associations were found.

On the other hand, we proposed simple ways to determine the longitudinal changes in the structural brain connectome using datasets of longitudinal diffusion weighted imaging and allele frequencies in single nucleotide polymorphisms (Elsheikh et al., 2020b), using data provided by the Alzheimer's Disease Neuroimaging Initiative (ADNI). We evaluated the contribution of genetic variations to the brain connectome changes in healthy individuals, early mild cognitive impaired and Alzheimer's patients. Specifically, we utilized the global connectivity metrics of the structural connectome, namely the segregation and integration features (Rubinov and Sporns, 2010), at two time points to quantify the longitudinal changes. We used these to carry out a GWAS and utilised the 1000 Genomes reference population to obtain the gene- and pathway-scores. We believe this analysis is the first of its type to associate the longitudinal shift in the structural connectome with whole genome variants besides our initial work in this book, which studies the correlation between gene expression and connectivity metrics (Elsheikh et al., 2019a). Our analysis success-fully identified genetic variants that associate with the structural connectivity changes in AD. Some of the identified variants have been previously reported to be associated with the risk of developing AD and other neurodegenerative diseases (De Ferrari et al., 2007; Kang et al., 2010; Young et al., 2012; Nicolas et al., 2013).

Aiming to understand the effects of multiple factors on the progression of AD, we also conducted an integrated association analysis of gene expression with longitudinal connectivity metrics and dementia severity measures. We examined these associations globally and regionally at distinct parts of the brain. These segments are based on the automated anatomical labeling segmentation (Elsheikh et al., 2018a). We retrieved the risk genes that are known for their manipulation of Alzheimer's risk and examined the relationship of their expression with the change in clinical dementia ratings and connectivity metrics. Moreover, using multivariate ridge regression, we studied the effect of brain connectivity changes and gene expression on dementia progression, using non-parametric associations and quantile regression. We show that expression of some gene, namely, *HFE*, *PLAU*, *BLMH* and *APP* affect the brain activities in the right anterior cingulate, Fusiform gyrus, Heschl gyrus and middle temporal gyrus in AD, respectively. Comparing these patterns of association with previous discoveries (Robson et al., 2004; Pujol et al., 2002; Finckh et al., 2003; Papassotiropoulos et al., 2000), our results suggest that the level at which these genes are expressed can cause changes in the pattern of connectivity in the AD brain.

To improve such studies we also developed a distribution-free, robust, fast and unbiased imaging genetics model, BiGen (Elsheikh et al., 2020a), using the idea of the partial least square (PLS) structural equation model (SEM) Lohmöller (1989). The code is written in Python and is made available and accessible online. BiGen integrates neuroimaging, multi-omic and clinical character-istics to analyse brain-related diseases. We applied the model to Alzheimer's dataset from ADNI repository and tested it on simulated data. We examined different hypotheses to understand the multivariate relationship and effect of genetic, endophenotypes - measured as brain connectivity metrics - and other confounding factos on the disease phenotypes. Here we used a four-step iterative algorithm to estimate the latent variables of the model, after which we estimated the inner path parameters using a quantile regression and included some interaction terms.

In conclusion, we proposed non-parametric analysis pipelines and developed a robust method us-ing the idea of SEM to study the effect of genetic variations on multi-dimensional neuroimaging phenotypes. We believe our findings and proposed model are useful in understanding the complex molecular heterogeneity underlying glioblastoma and AD. Consequently, this can facilitate a genetic diagnosis and personalized therapy. We also identified genes and variants that affect the connectivity at particular anatomical regions in the brain, which is useful in uncovering the effects of genetics on the brain connectivity structure for further improvement of patient care.

References

Hugo JWL Aerts. The potential of radiomic-based phenotyping in precision medicine: a review. *JAMA oncology*, 2(12):1636–1642, 2016.

Hugo JWL Aerts, Emmanuel Rios Velazquez, Ralph TH Leijenaar, Chintan Parmar, Patrick Grossmann, Sara Carvalho, Johan Bussink, René Monshouwer, Benjamin Haibe-Kains, Derek Rietveld, et al. Decoding tumour phenotype by noninvasive imaging using a quantitative radiomics approach. *Nature communications*, 5:4006, 2014.

Iman Aganj, Christophe Lenglet, Guillermo Sapiro, Essa Yacoub, Kamil Ugurbil, and Noam Harel. Reconstruction of the orientation distribution function in single-and multiple-shell q-ball imaging within constant solid angle. *Magnetic resonance in medicine*, 64(2):554–566, 2010.

Hamed Akbari, Spyridon Bakas, Jared M Pisapia, MacLean P Nasrallah, Martin Rozycki, Maria Martinez-Lage, Jennifer JD Morrissette, Nadia Dahmane, Donald M O'Rourke, and Christos Davatzikos. In vivo evaluation of egfrviii mutation in primary glioblastoma patients via complex multiparametric mri signature. *Neuro-oncology*, 2018.

Marilyn S Albert, Steven T DeKosky, Dennis Dickson, Bruno Dubois, Howard H Feldman, Nick C Fox, Anthony Gamst, David M Holtzman, William J Jagust, Ronald C Petersen, et al. The diagnosis of mild cognitive impairment due to Alzheimer's disease: Recommendations from the national institute on aging-Alzheimer's association workgroups on diagnostic guidelines for Alzheimer's disease. *Alzheimer's & dementia*, 7(3):270–279, 2011.

Andrew L Alexander, Jee Eun Lee, Mariana Lazar, and Aaron S Field. Diffusion tensor imaging of the brain. *Neurotherapeutics*, 4(3):316–329, 2007.

C Alloza, Simon R Cox, M Blesa Cábez, P Redmond, HC Whalley, SJ Ritchie, S Muñoz Maniega, M del C Valdés Hernández, EM Tucker-Drob, Stephen M Lawrie, et al. Polygenic risk score for schizophrenia and structural brain connectivity in older age: a longitudinal connectome and tractography study. *Neuroimage*, 183:884–896, 2018.

Moses Amadasun and Robert King. Textural features corresponding to textural properties. *IEEE Transactions on systems, man, and Cybernetics*, 19(5):1264–1274, 1989.

Christian Nicolaj Andreassen, Jan Alsner, and Jens Overgaard. Does variability in normal tissue reactions after radiotherapy have a genetic basis–where and how to look for it? *Radiotherapy and Oncology*, 64(2):131–140, 2002.

Henry Michael Arrighi, Peter J Neumann, Ivan M Lieberburg, and Raymond J Townsend. Lethality of Alzheimer disease and its impact on nursing home placement. *Alzheimer Disease & Associated Disorders*, 24(1):90–95, 2010.

Ya-hui Bai, Yun-bo Zhan, Bin Yu, Wei-Wei Wang, Li Wang, Jin-qiao Zhou, Ruo-kun Chen, Feng-jiang Zhang, Xin-wei Zhao, Wen-chao Duan, et al. A novel tumor-suppressor, cdh18, inhibits glioma cell invasiveness via UQCRC2 and correlates with the prognosis of glioma patients. *Cellular Physiology and Biochemistry*, 48(4):1755–1770, 2018.

S Bakas, H Akbari, A Sotiras, M Bilello, M Rozycki, J Kirby, J Freymann, K Farahani, and C Davatzikos. Segmentation labels and radiomic features for the pre-operative scans of the tcga-gbm collection. *The Cancer Imaging Archive*, 286, 2017a.

Spyridon Bakas, Ke Zeng, Aristeidis Sotiras, Saima Rathore, Hamed Akbari, Bilwaj Gaonkar, Martin Rozycki, Sarthak Pati, and Christos Davatzikos. Glistrboost: combining multimodal mri segmentation, registration, and biophysical tumor growth modeling with gradient boosting machines for glioma segmentation. In *International Workshop on Brainlesion: Glioma, Multiple Sclerosis, Stroke and Traumatic Brain Injuries*, pages 144–155. Springer, 2015.

Spyridon Bakas, Hamed Akbari, Jared Pisapia, Maria Martınez-Lage, Martin Rozycki, Saima Rathore, Nadia Dahmane, Donald M O'Rourke, and Christos Davatzikos. In vivo detection of egfrviii in glioblastoma via perfusion magnetic resonance imaging signature consistent with deep peritumoral infiltration: the φ-index. *Clinical Cancer Research*, 23(16):4724–4734, 2017b.

Spyridon Bakas, Hamed Akbari, Aristeidis Sotiras, Michel Bilello, Martin Rozycki, Justin S Kirby, John B Freymann, Keyvan Farahani, and Christos Davatzikos. Advancing the cancer genome atlas glioma mri collections with expert segmentation labels and radiomic features. *Scientific data*, 4:170117, 2017c.

Cliveand Ballard, Serge Gauthier, Corbett Anne, Carol Brayne, Dag Aarsland, and Emma Jones. Alzheimer's disease. *The Lancet*, 377(9770):1019 – 1031, 2011.

Deborah E Barnes and Kristine Yaffe. The projected effect of risk factor reduction on Alzheimer's disease prevalence. *The Lancet Neurology*, 10(9):819–828, 2011.

Nematollah K Batmanghelich, Adrian V Dalca, Mert R Sabuncu, and Polina Golland. Joint modeling of imaging and genetics. In *International Conference on Information Processing in Medical Imaging*, pages 766–777. Springer, 2013.

Johannes Bedenbender, Frieder M Paulus, Sören Krach, Martin Pyka, Jens Sommer, Axel Krug, Stephanie H Witt, Marcella Rietschel, Davide Laneri, Tilo Kircher, et al. Functional connectivity analyses in imaging genetics: considerations on methods and data interpretation. *PLoS One*, 6(12):e26354, 2011.

Lynn M Bekris, Chang-En Yu, Thomas D Bird, and Debby W Tsuang. Genetics of Alzheimer disease. *Journal of geriatric psychiatry and neurology*, 23(4):213–227, 2010.

L. Bertram and other. The genetics of Alzheimer disease: back to the future. *Neuron*, 68: 270–281, 2010.

Michel Bilello, Hamed Akbari, Xiao Da, Jared M Pisapia, Suyash Mohan, Ronald L Wolf, Donald M O'Rourke, Maria Martinez-Lage, and Christos Davatzikos. Population-based mri atlases of spatial distribution are specific to patient and tumor characteristics in glioblastoma. *NeuroImage: Clinical*, 12:34–40, 2016.

Zev A Binder, Amy Haseley Thorne, Spyridon Bakas, E Paul Wileyto, Michel Bilello, Hamed Akbari, Saima Rathore, Sung Min Ha, Logan Zhang, Cole J Ferguson, et al. Epidermal growth factor receptor extracellular domain mutations in glioblastoma present opportunities for clinical imaging and therapeutic development. *Cancer cell*, 34(1):163–177, 2018.

Kenneth A Bollen and J Scott Long. *Testing structural equation models*, volume 154. Sage, 1993.

Benjamin M BolstAD, Rafael A Irizarry, Magnus Åstrand, and Terence P. Speed. A comparison of normalization methods for high density oligonucleotide array data based on variance and bias. *Bioinformatics*, 19(2):185–193, 2003.

Heiko Braak, Dietmar R Thal, Estifanos Ghebremedhin, and Kelly Del Tredici. Stages of the pathologic process in Alzheimer disease: age categories from 1 to 100 years. *Journal of Neuropathology & Experimental Neurology*, 70(11):960–969, 2011.

Ron Brookmeyer, Maria M Corrada, Frank C Curriero, and Claudia Kawas. Survival following a diagnosis of Alzheimer disease. *Archives of neurology*, 59(11):1764–1767, 2002.

J.A. Brown, K.H. Terashima, A.C. Burggren, L.M. Ercoli, K.J. Miller, and other. Bain network local interconnectivity loss in aging apoe-4 allele carriers. *Proceedings of the National AcADemy of Sciences*, 108(51):20760–20765, 2011.

Brendan Bulik-Sullivan, Hilary K Finucane, Verneri Anttila, Alexander Gusev, Felix R Day, Po-Ru Loh, Laramie Duncan, John RB Perry, Nick Patterson, Elise B Robinson, et al. An atlas of genetic correlations across human diseases and traits. *Nature genetics*, 47(11):1236, 2015.

Patricia Oliveira Carminati, Stephano Spano Mello, Ana Lucia Fachin, Cristina Moraes Junta, Paula Sandrin-Garcia, Carlos Gilberto Carlotti, Eduardo Antonio Donadi, Geraldo Aleixo Silva Passos, and Elza Tiemi Sakamoto-Hojo. Alterations in gene expression profiles correlated with cisplatin cytotoxicity in the glioma u343 cell line. *Genetics and molecular biology*, 33(1): 159–168, 2010.

Christopher C Chang, Carson C Chow, Laurent CAM Tellier, Shashaank Vattikuti, Shaun M Purcell, and James J Lee. Second-generation PLINK: rising to the challenge of larger and richer datasets. *Gigascience*, 4(1):7, 2015.

Y. Chen, K. Chen, et al. Disrupted functional and structural networks in cognitively normal elderly subjects with the APOE ϵ4 allele. *Neuropsychopharmacology*, 40:1181, 2015.

Eric C Chi, Genevera I Allen, Hua Zhou, Omid Kohannim, Kenneth Lange, and Paul M Thompson. Imaging genetics via sparse canonical correlation analysis. In *2013 IEEE 10th International Symposium on Biomedical Imaging*, pages 740–743. IEEE, 2013.

Patrizia A Chiesa, Enrica Cavedo, Simone Lista, Paul M Thompson, Harald Hampel, Alzheimer Precision Medicine Initiative, et al. Revolution of resting-state functional neuroimaging genetics in Alzheimer's disease. *Trends in neurosciences*, 40(8):469–480, 2017.

Olivier L Chinot, Wolfgang Wick, Warren Mason, Roger Henriksson, Frank Saran, Ryo Nishikawa, Antoine F Carpentier, Khe Hoang-Xuan, Petr Kavan, Dana Cernea, et al. Bevacizumab plus radiotherapy–temozolomide for newly diagnosed glioblastoma. *New England Journal of Medicine*, 370(8):709–722, 2014.

Daniel Chow, Peter Chang, Brent D Weinberg, Daniela A Bota, Jack Grinband, and Christopher G Filippi. Imaging genetic heterogeneity in glioblastoma and other glial tumors: review of current methods and future directions. *American Journal of Roentgenology*, 210(1):30–38, 2018.

A Chu, Chandra M Sehgal, and James F Greenleaf. Use of gray value distribution of run lengths for texture analysis. *Pattern Recognition Letters*, 11(6):415–419, 1990.

Kenneth Clark, Bruce Vendt, Kirk Smith, John Freymann, Justin Kirby, Paul Koppel, Stephen Moore, Stanley Phillips, David Maffitt, Michael Pringle, et al. The cancer imaging archive (tcia): maintaining and operating a public information repository. *Journal of digital imaging*, 26(6):1045–1057, 2013.

1000 Genomes Project Consortium et al. An integrated map of genetic variation from 1,092 human genomes. *Nature*, 491(7422):56, 2012.

E. Corder, A. Saunders, et al. Gene dose of apolipoprotein E type 4 allele and the risk of Alzheimer's disease in late onset families. *Science*, 261(5123):921–923, 1993.

EH Corder, Al M Saunders, NJ Risch, WJ Strittmatter, DE Schmechel, PC Gaskell, JB Rimmler, PA Locke, PM Conneally, KE Schmader, et al. Protective effect of apolipoprotein e type 2 allele for late onset Alzheimer disease. *Nature genetics*, 7(2):180, 1994.

Elise Cuyvers and Kristel Sleegers. Genetic variations underlying Alzheimer's disease: evidence from genome-wide association studies and beyond. *The Lancet Neurology*, 15(8):857–868, 2016.

M. Daianu, A. Mezher, N. Jahanshad, D.P. Hibar, T.M. Nir, et al. Spectral graph theory and graph energy metrics show evidence for the Alzheimer's disease disconnection syndrome in APOE-4 risk gene carriers. In *Biomedical Imaging, 2015 IEEE 12th International Symposium on*, pages 458–461, 2015.

Belur V Dasarathy and Edwin B Holder. Image characterizations based on joint gray level—run length distributions. *Pattern Recognition Letters*, 12(8):497–502, 1991.

Christos Davatzikos, Saima Rathore, Spyridon Bakas, Sarthak Pati, Mark Bergman, Ratheesh Kalarot, Patmaa Sridharan, Aimilia Gastounioti, Nariman Jahani, Eric Cohen, et al. Cancer imaging phenomics toolkit: quantitative imaging analytics for precision diagnostics and predictive modeling of clinical outcome. *Journal of Medical Imaging*, 5(1):011018, 2018.

Mary Elizabeth Davis. Glioblastoma: overview of disease and treatment. *Clinical journal of oncology nursing*, 20(5):S2, 2016.

Giancarlo V De Ferrari, Andreas Papassotiropoulos, Travis Biechele, Fabienne Wavrant De-Vrieze, Miguel E Avila, Michael B Major, Amanda Myers, Katia Sáez, Juan P Henríquez, Alice Zhao, et al. Common genetic variation within the low-density lipoprotein receptor-related protein 6 and late-onset Alzheimer's disease. *Proceedings of the National AcADemy of Sciences*, 104 (22):9434–9439, 2007.

G. Deco, G. Tononi, et al. Rethinking segregation and integration: contributions of whole-brain modelling. *Nature Reviews Neuroscience*, 16:430–439, 2015.

Maximilian Diehn, Christine Nardini, David S Wang, Susan McGovern, Mahesh Jayaraman, Yu Liang, Kenneth Aldape, Soonmee Cha, and Michael D Kuo. Identification of noninvasive imaging surrogates for brain tumor gene-expression modules. *Proceedings of the National Academy of Sciences*, 105(13):5213–5218, 2008.

Stephen S Dominy, Casey Lynch, Florian Ermini, Malgorzata Benedyk, Agata Marczyk, Andrei KonrADi, Mai Nguyen, Ursula HADitsch, Debasish Raha, Christina Griffin, et al. Porphyromonas gingivalis in Alzheimer's disease brains: Evidence for disease causation and treatment with small-molecule inhibitors. *Science ADvances*, 5(1):eaau3333, 2019.

Gwenaëlle Douaud, Stephen Smith, Mark Jenkinson, Timothy Behrens, Heidi Johansen-Berg, John Vickers, Susan James, Natalie Voets, Kate Watkins, Paul M Matthews, et al. Anatomically related grey and white matter abnormalities in adolescent-onset schizophrenia. *Brain*, 130(9):2375–2386, 2007.

Oonagh Dowling, Analisa Difeo, Maria C Ramirez, Turgut Tukel, Goutham Narla, Luisa Bonafe, Hulya Kayserili, Memnune Yuksel-Apak, Amy S Paller, Karen Norton, et al. Mutations in capillary morphogenesis gene-2 result in the allelic disorders juvenile hyaline fibromatosis and infantile systemic hyalinosis. *The American Journal of Human Genetics*, 73(4):957–966, 2003.

Sylvia Drabycz, Gloria Roldán, Paula De Robles, Daniel Adler, John B McIntyre, Anthony M Magliocco, J Gregory Cairncross, and J Ross Mitchell. An analysis of image texture, tumor location, and mgmt promoter methylation in glioblastoma using magnetic resonance imaging. *Neuroimage*, 49(2):1398–1405, 2010.

H Duffau. Awake mapping of the brain connectome in glioma surgery: concept is stronger than technology. *European Journal of Surgical Oncology*, 41(9):1261–1263, 2015.

Véronique Duhem-Tonnelle, Ivan Bièche, Sophie Vacher, Anne Loyens, Claude-Alain Maurage, Francis Collier, Marc Baroncini, Serge Blond, Vincent Prevot, and Ariane Sharif. Differential distribution of erbb receptors in human glioblastoma multiforme: expression of erbb3 in cd133-positive putative cancer stem cells. *Journal of Neuropathology & Experimental Neurology*, 69 (6):606–622, 2010.

Team Egs. Enigma2 1kgp cookbook (v3). *Enhancing Neuroimaging Genetics through MetaAnalysis (ENIGMA) Consortium*, 2013.

Elizabeth A Eisenhauer, Patrick Therasse, Jan Bogaerts, Lawrence H Schwartz, D Sargent, Robert Ford, Janet Dancey, S Arbuck, Steve Gwyther, Margaret Mooney, et al. New response evaluation criteria in solid tumours: revised recist guideline (version 1.1). *European journal of cancer*, 45(2):228–247, 2009.

Benjamin M Ellingson. Radiogenomics and imaging phenotypes in glioblastoma: novel observations and correlation with molecular characteristics. *Current neurology and neuroscience reports*, 15(1):506, 2015.

Benjamin M Ellingson, Patrick Y Wen, Martin J van den Bent, and Timothy F Cloughesy. Pros and cons of current brain tumor imaging. *Neuro-oncology*, 16(suppl_7):vii2–vii11, 2014.

BM Ellingson, A Lai, RJ Harris, JM Selfridge, WH Yong, K Das, WB Pope, PL Nghiemphu, HV Vinters, LM Liau, et al. Probabilistic radiographic atlas of glioblastoma phenotypes. *American Journal of neuroradiology*, 2012.

Susan V Ellor, Teri Ann Pagano-Young, and Nicholas G Avgeropoulos. Glioblastoma: background, standard treatment paradigms, and supportive care considerations, 2014.

S Elsheikh, Emile R Chimusa, N Mulder, and A Crimi. Relating connectivity changes in brain networks to genetic information in Alzheimer patients. In *2018 IEEE 15th International Symposium on Biomedical Imaging (ISBI 2018)*, pages 1390–1393. IEEE, 2018a.

Samar S. M. Elsheikh, Emile R. Chimusa, Nicola Mulder, and Alessandro Crimi. Relating global and local connectome changes to dementia and targeted gene expression in Alzheimer's disease. *bioRxiv*, 2019a. doi: 10.1101/730416. URL https://www.biorxiv.org/content/early/2019/08/08/730416.

Samar S. M. Elsheikh, Emile R. Chimusa, Nicola J. Mulder, and Alessandro Crimi. Genome-wide association study of brain connectivity changes for Alzheimer's disease. *bioRxiv*, 2019b. doi: 10.1101/342436. URL https://www.biorxiv.org/content/early/2019/05/30/342436.

Samar S. M. Elsheikh, Emile R. Chimusa, Alessandro Crimi, and Nicola J. Mulder. Bigen: Integrative clinical and brain-imaging genetics analysis using structural equation model. *bioRxiv*, 2020a. doi: 10.1101/2020.02.04.934596. URL https://www.biorxiv.org/content/early/2020/02/05/2020.02.04.934596.

Samar SM Elsheikh, Spyridon Bakas, Nicola J Mulder, Emile R Chimusa, Christos Davatzikos, and Alessandro Crimi. Multi-stage association analysis of glioblastoma gene expressions with texture and spatial patterns. In *International MICCAI Brainlesion Workshop*, pages 239–250. Springer, 2018b.

Samar SM Elsheikh, Emile R Chimusa, Nicola J Mulder, and Alessandro Crimi. Genome-wide association study of brain connectivity changes for alzheimer's disease. *Scientific Reports*, 10 (1):1–16, 2020b.

Valentina Escott-Price, Céline Bellenguez, Li-San Wang, Seung-Hoan Choi, Denise Harold, Lesley Jones, Peter Holmans, Amy Gerrish, Alexey Vedernikov, Alexander Richards, et al. Gene-wide analysis detects two new susceptibility genes for Alzheimer's disease. *PloS one*, 9(6):e94661, 2014.

Linn Fagerberg, Björn M Hallström, Per Oksvold, Caroline Kampf, Dijana Djureinovic, Jacob Odeberg, Masato Habuka, Simin Tahmasebpoor, Angelika Danielsson, Karolina Edlund, et al. Analysis of the human tissue-specific expression by genome-wide integration of transcriptomics and antibody-based proteomics. *Molecular & Cellular Proteomics*, 13(2):397–406, 2014.

Jian Fang, Dongdong Lin, S Charles Schulz, Zongben Xu, Vince D Calhoun, and Yu-Ping Wang. Joint sparse canonical correlation analysis for detecting differential imaging genetics modules. *Bioinformatics*, 32(22):3480–3488, 2016.

Nicolas Farina, Jennifer Rusted, and Naji Tabet. The effect of exercise interventions on cognitive outcome in Alzheimer's disease: a systematic review. *International Psychogeriatrics*, 26(1): 9–18, 2014.

Lindsay A Farrer, Carmela R Abraham, Jonathan L Haines, Ekaterina A Rogaeva, Youqiang Song, Walker T McGraw, Nicholas Brindle, Smita Premkumar, William K Scott, Larry H Yamaoka, et al. Association between bleomycin hydrolase and Alzheimer's disease in caucasians. *Annals of neurology*, 44(5):808–811, 1998.

U Finckh, K Van Hadeln, T Müller-Thomsen, A Alberici, G Binetti, C Hock, RM Nitsch, G Stoppe, J Reiss, and A Gal. Association of late-onset Alzheimer disease with a genotype of plau, the gene encoding urokinase-type plasminogen activator on chromosome 10q22. 2. *Neurogenetics*, 4(4):213–217, 2003.

Hilary K Finucane, Brendan Bulik-Sullivan, Alexander Gusev, Gosia Trynka, Yakir Reshef, Po-Ru Loh, Verneri Anttila, Han Xu, Chongzhi Zang, Kyle Farh, et al. Partitioning heritability by functional annotation using genome-wide association summary statistics. *Nature genetics*, 47 (11):1228, 2015.

Centers for Disease Control, Prevention, et al. About underlying cause of death, 1999-2017, 2018.

Lars Forsberg, Sigurdur Sigurdsson, Lenore J Launer, Vilmundur Gudnason, and Fredrik Ullén. Structural covariability hubs in old age. *NeuroImage*, 189:307–315, 2019.

Laura Fratiglioni, Anders Ahlbom, Matti Viitanen, and Bengt Winblad. Risk factors for late-onset Alzheimer's disease: A population-based, case-control study. *Annals of neurology*, 33 (3):258–266, 1993.

Chris Gaiteri, Sara Mostafavi, Christopher J Honey, Philip L De Jager, and David A Bennett. Genetic variants in Alzheimer disease—molecular and brain network approaches. *Nature Reviews Neurology*, 12(7):413, 2016.

Mary M Galloway. Texture analysis using grey level run lengths. *NASA STI/Recon Technical Report N*, 75, 1974.

Clare J Galton, Karalyn Patterson, K Graham, Matthew A Lambon-Ralph, G Williams, N Antoun, BJ Sahakian, and JR Hodges. Differing patterns of temporal atrophy in Alzheimer's disease and semantic dementia. *Neurology*, 57(2):216–225, 2001.

Eleftherios Garyfallidis, Matthew Brett, Bago Amirbekian, Ariel Rokem, Stéfan van der Walt, Maxime Descoteaux, and Ian Nimmo-Smith. Dipy, a library for the analysis of diffusion MRI data. *Frontiers in Neuroinformatics*, 8(8), 2014. doi: 10.3389/fninf.2014.00008. URL http://dx.doi.org/10.3389/fninf.2014.00008.

Margaret Gatz, Nancy L Pedersen, Stig Berg, Boo Johansson, Kurt Johansson, James A Mortimer, Samuel F Posner, Matti Viitanen, Bengt Winblad, and Anders Ahlbom. Heritability for Alzheimer's disease: the study of dementia in swedish twins. *The Journals of Gerontology Series A: Biological Sciences and Medical Sciences*, 52(2):M117–M125, 1997.

Joseph Gaugler, Bryan James, Tricia Johnson, Allison Marin, and Jennifer Weuve. 2019 Alzheimer's disease facts and figures. *Alzheimers & Dementia*, 15(3):321–387, 2019.

Alison Goate, Marie-Christine Chartier-Harlin, Mike Mullan, Jeremy Brown, Fiona Crawford, Liana Fidani, Luis Giuffra, Andrew Haynes, Nick Irving, Louise James, et al. Segregation of a missense mutation in the amyloid precursor protein gene with familial Alzheimer's disease. *Nature*, 349(6311):704, 1991.

Christopher G Goetz. *Textbook of clinical neurology*, volume 355. Elsevier Health Sciences, 2007.

Jill S Goldman, Susan E Hahn, Jennifer Williamson Catania, Susan Larusse-Eckert, Melissa Barber Butson, Malia Rumbaugh, Michelle N Strecker, J Scott Roberts, Wylie Burke, Richard Mayeux, et al. Genetic counseling and testing for Alzheimer disease: joint practice guidelines of the american college of medical genetics and the national society of genetic counselors. *Genetics in Medicine*, 13(6):597, 2011.

Catriona D Good, Ingrid S Johnsrude, John Ashburner, Richard NA Henson, Karl J Friston, and Richard SJ Frackowiak. A voxel-based morphometric study of ageing in 465 normal adult human brains. *Neuroimage*, 14(1):21–36, 2001.

Ali Gooya, Kilian M Pohl, Michel Bilello, Luigi Cirillo, George Biros, Elias R Melhem, and Christos Davatzikos. Glistr: glioma image segmentation and registration. *IEEE transactions on medical imaging*, 31(10):1941–1954, 2012.

Brian A Gordon, Tyler M Blazey, Yi Su, Amrita Hari-Raj, Aylin Dincer, Shaney Flores, Jon Christensen, Eric McDade, Guoqiao Wang, Chengjie Xiong, et al. Spatial patterns of neuroimaging biomarker change in individuals from families with autosomal dominant Alzheimer's disease: a longitudinal study. *The Lancet Neurology*, 17(3):241–250, 2018.

Evan M Gordon, Timothy O Laumann, Babatunde ADeyemo, Jeremy F Huckins, William M Kelley, and Steven E Petersen. Generation and evaluation of a cortical area parcellation from resting-state correlations. *Cerebral cortex*, 26(1):288–303, 2014.

Andrew D Grotzinger, Mijke Rhemtulla, Ronald de Vlaming, Stuart J Ritchie, Travis T Mallard, W David Hill, Hill F Ip, Andrew M McIntosh, Ian J Deary, Philipp D Koellinger, et al. Genomic sem provides insights into the multivariate genetic architecture of complex traits. *BioRxiv*, page 305029, 2018.

David A Gutman, Lee AD Cooper, Scott N Hwang, Chad A Holder, JingJing Gao, Tarun D Aurora, William D Dunn Jr, Lisa Scarpace, Tom Mikkelsen, Rajan Jain, et al. Mr imaging predictors of molecular profile and survival: multi-institutional study of the tcga glioblastoma data set. *Radiology*, 267(2):560–569, 2013.

Patric Hagmann, Leila Cammoun, Xavier Gigandet, Reto Meuli, Christopher J Honey, Van J Wedeen, and Olaf Sporns. Mapping the structural core of human cerebral cortex. *PLoS biology*, 6(7):e159, 2008.

D Hamer and Leo Sirota. Beware the chopsticks gene, 2000.

Farina Hanif, Kanza Muzaffar, Kahkashan Perveen, Saima M Malhi, and Shabana U Simjee. Glioblastoma multiforme: a review of its epidemiology and pathogenesis through clinical presentation and treatment. *Asian Pacific journal of cancer prevention: APJCP*, 18(1):3, 2017.

Sandra Hanks, Sarah ADams, Jenny Douglas, Laura Arbour, David J Atherton, Sevim Balci, Harald Bode, Mary E Campbell, Murray Feingold, Gökhan Keser, et al. Mutations in the gene encoding capillary morphogenesis protein 2 cause juvenile hyaline fibromatosis and infantile systemic hyalinosis. *The American Journal of Human Genetics*, 73(4):791–800, 2003.

Robert M Haralick, Karthikeyan Shanmugam, et al. Textural features for image classification. *IEEE Transactions on systems, man, and cybernetics*, (6):610–621, 1973.

Michael Hart, Stephen J Price, and John Suckling. 141 preoperative brain mapping in neuro-oncology with graph theory analysis of the functional connectome. *Neurosurgery*, 62 (CN_suppl_1):211–211, 2015.

Peng He, Lin Sun, Dan Zhu, Hong Zhang, Liang Zhang, Yujie Guo, Siwen Liu, Jingjing Zhou, Xiaoyan Xu, and Peng Xie. Knock-down of endogenous bornavirus-like nucleoprotein 1 inhibits cell growth and induces apoptosis in human oligodendroglia cells. *International journal of molecular sciences*, 17(4):435, 2016.

LE Hebert, JL Bienias, NT Aggarwal, RS Wilson, DA Bennett, RC Shah, and DA Evans. Change in risk of Alzheimer disease over time. *Neurology*, 75(9):786–791, 2010.

Monika E Hegi, Annie-Claire Diserens, Thierry Gorlia, Marie-France Hamou, Nicolas De Tribolet, Michael Weller, Johan M Kros, Johannes A Hainfellner, Warren Mason, Luigi Mariani, et al. Mgmt gene silencing and benefit from temozolomide in glioblastoma. *New England Journal of Medicine*, 352(10):997–1003, 2005.

Elizabeth P Helzner, Nikolaos Scarmeas, Stephanie Cosentino, MX Tang, Nicole Schupf, and Yaakov Stern. Survival in Alzheimer disease: a multiethnic, population-based study of incident cases. *Neurology*, 71(19):1489–1495, 2008.

Jörg Henseler, Christian M Ringle, and Marko Sarstedt. Using partial least squares path modeling in advertising research: basic concepts and recent issues. *Handbook of research on international advertising*, 252, 2012.

Suzana Herculano-Houzel. The human brain in numbers: a linearly scaled-up primate brain. *Frontiers in human neuroscience*, 3:31, 2009.

Derrek P Hibar, Jason L Stein, Miguel E Renteria, Alejandro Arias-Vasquez, Sylvane Desrivières, Neda JahanshAD, Roberto Toro, Katharina Wittfeld, Lucija Abramovic, Micael Andersson, et al. Common genetic variants influence human subcortical brain structures. *Nature*, 520 (7546):224, 2015.

Joel N Hirschhorn and Mark J Daly. Genome-wide association studies for common diseases and complex traits. *Nature reviews genetics*, 6(2):95, 2005.

Dominique B Hoelzinger, Luigi Mariani, Joachim Weis, Tanja Woyke, Theresa J Berens, Wendy S McDonough, Andrew Sloan, Stephen W Coons, and Michael E Berens. Gene expression profile of glioblastoma multiforme invasive phenotype points to new therapeutic targets. *Neoplasia (New York, NY)*, 7(1):7, 2005.

Arthur E Hoerl and Robert W Kennard. Ridge regression: Biased estimation for nonorthogonal problems. *Technometrics*, 12(1):55–67, 1970.

Cosmina Hogea, George Biros, Feby Abraham, and Christos Davatzikos. A robust framework for soft tissue simulations with application to modeling brain tumor mass effect in 3d mr images. *Physics in Medicine & Biology*, 52(23):6893, 2007a.

Cosmina Hogea, Christos Davatzikos, and George Biros. Modeling glioma growth and mass effect in 3d mr images of the brain. In *International Conference on Medical Image Computing and Computer-Assisted Intervention*, pages 642–650. Springer, 2007b.

Cosmina Hogea, Christos Davatzikos, and George Biros. An image-driven parameter estimation problem for a reaction–diffusion glioma growth model with mass effects. *Journal of mathematical biology*, 56(6):793–825, 2008.

Raymond Y Huang, Martha R Neagu, David A Reardon, and Patrick Y Wen. Pitfalls in the neuroimaging of glioblastoma in the era of antiangiogenic and immuno/targeted therapy-detecting illusive disease, defining response. *Frontiers in neurology*, 6:33, 2015.

Sjoerd MH Huisman, Ahmed Mahfouz, Nematollah K Batmanghelich, Boudewijn PF Lelieveldt, Marcel JT Reinders, Alzheimer's Disease Neuroimaging Initiative, et al. A structural equation model for imaging genetics using spatial transcriptomics. *Brain informatics*, 5(2):13, 2018.

Gabriel Iacob and Eduard B Dinca. Current data and strategy in glioblastoma multiforme. *Journal of medicine and life*, 2(4):386, 2009.

R. Irizarry et al. Exploration, normalization, and summaries of high density oligonucleotide array probe level data. *Biostatistics*, 4:249–264, 2003.

Haruka Itakura, Achal S Achrol, Lex A Mitchell, Joshua J Loya, Tiffany Liu, Erick M Westbroek, Abdullah H Feroze, Scott Rodriguez, Sebastian Echegaray, Tej D Azad, et al. Magnetic resonance image features identify glioblastoma phenotypic subtypes with distinct molecular pathway activities. *Science translational medicine*, 7(303):303ra138–303ra138, 2015.

Clifford R Jack Jr, Matt A Bernstein, Nick C Fox, Paul Thompson, Gene Alexander, Danielle Harvey, Bret Borowski, Paula J Britson, Jennifer L. Whitwell, ChADwick Ward, et al. The Alzheimer's disease neuroimaging initiative (ADNI): MRI methods. *Journal of Magnetic Resonance Imaging: An Official Journal of the International Society for Magnetic Resonance in Medicine*, 27(4):685–691, 2008.

Neda JahanshAD, Priya Rajagopalan, Xue Hua, Derrek P Hibar, Talia M Nir, Arthur W Toga, Clifford R Jack, Andrew J Saykin, Robert C Green, Michael W Weiner, et al. Genome-wide scan of healthy human connectome discovers SPON1 gene variant influencing dementia severity. *Proceedings of the National AcADemy of Sciences*, 110(12):4768–4773, 2013.

Akane Kajiya, Hiroyuki Kaji, Toshiaki Isobe, and Atsushi Takeda. Processing of amyloid β-peptides by neutral cysteine protease bleomycin hydrolase. *Protein and peptide letters*, 13(2): 119–123, 2006.

Yeon-Joo Kang, Murat Digicaylioglu, Rossella Russo, Marcus Kaul, Cristian L Achim, Lauren Fletcher, Eliezer Masliah, and Stuart A Lipton. Erythropoietin plus insulin-like growth factor-i protects against neuronal damage in a murine model of human immunodeficiency virus-associated neurocognitive disorders. *Annals of neurology*, 68(3):342–352, 2010.

K Kantarci, SD Weigand, SA Przybelski, MM Shiung, Jennifer Lynn Whitwell, S Negash, David S Knopman, Bradley F Boeve, PC O'Brien, Ronald Carl Petersen, et al. Risk of dementia in mci: combined effect of cerebrovascular disease, volumetric mri, and 1h mrs. *Neurology*, 72(17): 1519–1525, 2009.

Maurice George Kendall et al. The advanced theory of statistics. vols. 1. *The advanced theory of statistics. Vols. 1.*, 1(Ed. 4), 1948.

Rhoda J Kinsella, Andreas Kähäri, Syed Haider, Jorge Zamora, Glenn Proctor, Giulietta Spudich, Jeff Almeida-King, Daniel Staines, Paul Derwent, Arnaud Kerhornou, et al. Ensembl biomarts: a hub for data retrieval across taxonomic space. *Database*, 2011, 2011.

Takahide Kodama, Eiji Ikeda, Aiko OkADa, Takashi Ohtsuka, Masayuki Shimoda, Takayuki Shiomi, Kazunari Yoshida, Mitsutoshi NakADa, Eiko Ohuchi, and Yasunori OkADa. Adam12 is selectively overexpressed in human glioblastomas and is associated with glioblastoma cell proliferation and shedding of heparin-binding epidermal growth factor. *The American journal of pathology*, 165(5):1743–1753, 2004.

Roger Koenker and Kevin F Hallock. Quantile regression. *Journal of economic perspectives*, 15 (4):143–156, 2001.

Jean-Charles Lambert, Carla A Ibrahim-Verbaas, Denise Harold, ADam C Naj, Rebecca Sims, Céline Bellenguez, Gyungah Jun, Anita L DeStefano, Joshua C Bis, Gary W Beecham, et al. Meta-analysis of 74,046 individuals identifies 11 new susceptibility loci for Alzheimer's disease. *Nature genetics*, 45(12):1452, 2013.

Philippe Lambin, Emmanuel Rios-Velazquez, Ralph Leijenaar, Sara Carvalho, Ruud GPM van Stiphout, Patrick Granton, Catharina ML Zegers, Robert Gillies, Ronald Boellard, André Dekker, et al. Radiomics: extracting more information from medical images using advanced feature analysis. *European journal of cancer*, 48(4):441–446, 2012.

David Lamparter, Daniel Marbach, Rico Rueedi, Zoltán Kutalik, and Sven Bergmann. Fast and rigorous computation of gene and pathway scores from SNP-based summary statistics. *PLoS computational biology*, 12(1):e1004714, 2016.

Suvi Larjavaara, Riitta Mäntylä, Tiina Salminen, Hannu Haapasalo, Jani Raitanen, Juha Jääskeläinen, and Anssi Auvinen. Incidence of gliomas by anatomic location. *Neuro-oncology*, 9(3):319–325, 2007.

Édith Le Floch, Vincent Guillemot, Vincent Frouin, Philippe Pinel, Christophe Lalanne, Laura Trinchera, Arthur Tenenhaus, Antonio Moreno, Monica Zilbovicius, Thomas Bourgeron, et al. Significant correlation between a set of genetic polymorphisms and a functional brain network revealed by feature selection and sparse partial least squares. *Neuroimage*, 63(1):11–24, 2012.

Ephrat Levy-Lahad, Wilma Wasco, Parvoneh Poorkaj, Donna M Romano, Junko Oshima, Warren H Pettingell, Chang-en Yu, Paul D Jondro, Stephen D Schmidt, Kai Wang, et al. Candidate gene for the chromosome 1 familial Alzheimer's disease locus. *Science*, 269(5226):973–977, 1995.

Hao Li, Sally Wetten, Li Li, Pamela L St Jean, Ruchi Upmanyu, Linda Surh, David Hosford, Michael R Barnes, James David Briley, Michael Borrie, et al. Candidate single-nucleotide polymorphisms from a genomewide association study of Alzheimer disease. *Archives of neurology*, 65(1):45–53, 2008.

Jingyu Liu and Vince D Calhoun. A review of multivariate analyses in imaging genetics. *Frontiers in neuroinformatics*, 8:29, 2014.

Jingyu Liu, Godfrey Pearlson, Andreas Windemuth, Gualberto Ruano, Nora I Perrone-Bizzozero, and Vince Calhoun. Combining fmri and snp data to investigate connections between brain function and genetics using parallel ica. *Human brain mapping*, 30(1):241–255, 2009.

Yun Liu, Lianrong Yu, Xiaoyan Wu, Yanli Chen, Ji Ge, Qun Li, and Zhongyuan Xue. Silencing cct6a suppresses cell migration and invasion in glioblastoma in vitro. *INTERNATIONAL JOURNAL OF CLINICAL AND EXPERIMENTAL MEDICINE*, 10(9):13263–13271, 2017.

Jan-Bernd Lohmöller. Predictive vs. structural modeling: Pls vs. ml. In *Latent variable path modeling with partial least squares*, pages 199–226. Springer, 1989.

David N Louis, Hiroko Ohgaki, Otmar D Wiestler, Webster K Cavenee, Peter C Burger, Anne Jouvet, Bernd W Scheithauer, and Paul Kleihues. The 2007 who classification of tumours of the central nervous system. *Acta neuropathologica*, 114(2):97–109, 2007.

Zhao-Hua Lu, Zakaria Khondker, Joseph G Ibrahim, Yue Wang, Hongtu Zhu, Alzheimer's Disease Neuroimaging Initiative, et al. Bayesian longitudinal low-rank regression models for imaging genetic data from longitudinal studies. *NeuroImage*, 149:305–322, 2017.

Marc C Mabray, Ramon F Barajas, and Soonmee Cha. Modern brain tumor imaging. *Brain tumor research and treatment*, 3(1):8–23, 2015.

David R Macdonald, Terrance L Cascino, S Clifford Schold Jr, J Gregory Cairncross, et al. Response criteria for phase ii studies of supratentorial malignant glioma. *J Clin Oncol*, 8(7): 1277–1280, 1990.

Luke Macyszyn, Hamed Akbari, Jared M Pisapia, Xiao Da, Mark Attiah, Vadim Pigrish, Yingtao Bi, Sharmistha Pal, Ramana V Davuluri, Laura Roccograndi, et al. Imaging patterns predict patient survival and molecular subtype in glioblastoma via machine learning techniques. *Neuro-oncology*, 18(3):417–425, 2015.

Teri A Manolio, Francis S Collins, Nancy J Cox, David B Goldstein, Lucia A Hindorff, David J Hunter, Mark I McCarthy, Erin M Ramos, Lon R Cardon, Aravinda Chakravarti, et al. Finding the missing heritability of complex diseases. *Nature*, 461(7265):747, 2009.

Urko M Marigorta, Juan Antonio Rodríguez, Greg Gibson, and Arcadi Navarro. Replicability and prediction: lessons and challenges from gwas. *Trends in Genetics*, 34(7):504–517, 2018.

W Ian McDonald, Alistair Compston, Gilles Edan, Donald Goodkin, Hans-Peter Hartung, Fred D Lublin, Henry F McFarland, Donald W Paty, Chris H Polman, Stephen C Reingold, et al. Recommended diagnostic criteria for multiple sclerosis: guidelines from the international panel on the diagnosis of multiple sclerosis. *Annals of Neurology: Official Journal of the American Neurological Association and the Child Neurology Society*, 50(1):121–127, 2001.

Guy M McKhann, David S Knopman, Howard Chertkow, Bradley T Hyman, Clifford R Jack Jr, Claudia H Kawas, William E Klunk, Walter J Koroshetz, Jennifer J Manly, Richard Mayeux, et al. The diagnosis of dementia due to Alzheimer's disease: Recommendations from the national institute on aging-Alzheimer's association workgroups on diagnostic guidelines for Alzheimer's disease. *Alzheimer's & dementia*, 7(3):263–269, 2011.

Sarah E Medland, Neda Jahanshad, Benjamin M Neale, and Paul M Thompson. Whole-genome analyses of whole-brain data: working within an expanded search space. *Nature neuroscience*, 17(6):791, 2014.

Walter Mier and Daniela Mier. Advantages in functional imaging of the brain. *Frontiers in human neuroscience*, 9:249, 2015.

Chie Mizumaru, Yuhki Saito, Takao Ishikawa, Tomohiro Yoshida, Tohru Yamamoto, TADashi Nakaya, and Toshiharu Suzuki. Suppression of APP-containing vesicle trafficking and production of β-amyloid by aid/dhhc-12 protein. *Journal of neurochemistry*, 111(5):1213–1224, 2009.

John C Morris, Catherine M Roe, Chengjie Xiong, Anne M Fagan, Alison M Goate, David M Holtzman, and Mark A Mintun. Apoe predicts amyloid-beta but not tau Alzheimer pathology in cognitively normal aging. *Annals of neurology*, 67(1):122–131, 2010.

John Carl Morris, Christopher Ernesto, Kimberly Schafer, MMSN Coats, SMSN Leon, M Sano, LJ Thal, and P Woodbury. Clinical dementia rating training and reliability in multicenter studies: the Alzheimer's disease cooperative study experience. *Neurology*, 48(6):1508–1510, 1997.

Maciej M Mrugala. Advances and challenges in the treatment of glioblastoma: a clinician's perspective. *Discovery medicine*, 15(83):221–230, 2013.

Sherry L Murphy, Jiaquan Xu, Kenneth D Kochanek, and Elizabeth Arias. Mortality in the united states, 2017. 2018.

ADam C Naj, Gary W Beecham, Eden R Martin, Paul J Gallins, Eric H Powell, Ioanna Konidari, Patrice L WhiteheAD, Guiqing Cai, Vahram Haroutunian, William K Scott, et al. Dementia revealed: novel chromosome 6 locus for late-onset Alzheimer disease provides genetic evidence for folate-pathway abnormalities. *PLoS genetics*, 6(9):e1001130, 2010.

Yoshio Namba, Yasuyoshi Ouchi, Atsushi Takeda, Akira Ueki, and Kazuhiko Ikeda. Bleomycin hydrolase immunoreactivity in senile plaque in the brains of patients with Alzheimer's disease. *Brain research*, 830(1):200–202, 1999.

Shawn R Narum. Beyond bonferroni: less conservative analyses for conservation genetics. *Conservation genetics*, 7(5):783–787, 2006.

National Institutes of Health. National Institute on Aging. What happens to the brain in Alzheimer's disease? https://www.nia.nih.gov/health/what-happens-brain-alzheimers-disease.

Anupma Nayak, Angela Mercy Ralte, Mehar Chand Sharma, Varinder Paul Singh, Ashok Kumar Mahapatra, Veer Singh Mehta, and Chitra Sarkar. p53 protein alterations in adult astrocytic tumors and oligodendrogliomas. *Neurology India*, 52(2):228, 2004.

Cancer Genome Atlas Research Network. Comprehensive, integrative genomic analysis of diffuse lower-grade gliomas. *New England Journal of Medicine*, 372(26):2481–2498, 2015.

Cancer Genome Atlas Research Network et al. Comprehensive genomic characterization defines human glioblastoma genes and core pathways. *Nature*, 455(7216):1061, 2008.

Celine S Nicolas, Mascia Amici, Zuner A Bortolotto, Andrew Doherty, Zsolt Csaba, Assia Fafouri, Pascal Dournaud, Pierre Gressens, Graham L Collingridge, and Stephane Peineau. The role of jak-stat signaling within the CNS. *Jak-Stat*, 2(1):e22925, 2013.

Sergey V Nuzhdin, Maren L Friesen, and Lauren M McIntyre. Genotype–phenotype mapping in a post-gwas world. *Trends in Genetics*, 28(9):421–426, 2012.

Hiroko Ohgaki and Paul Kleihues. Epidemiology and etiology of gliomas. *Acta neuropathologica*, 109(1):93–108, 2005.

Nuala A O'Leary, Mathew W Wright, J Rodney Brister, Stacy Ciufo, Diana HADdAD, Rich McVeigh, Bhanu Rajput, Barbara Robbertse, Brian Smith-White, Danso Ako-ADjei, et al. Reference sequence (refseq) database at ncbi: current status, taxonomic expansion, and functional annotation. *Nucleic acids research*, 44(D1):D733–D745, 2015.

Marie Orre, Willem Kamphuis, Stephanie Dooves, Lieneke Kooijman, Elena T Chan, Christopher J Kirk, Vanessa Dimayuga Smith, Sanne Koot, Carlyn Mamber, Anne H Jansen, et al. Reactive glia show increased immunoproteasome activity in Alzheimer's disease. *Brain*, 136(5):1415–1431, 2013.

A Papassotiropoulos, M Bagli, F Jessen, C Frahnert, ML Rao, W Maier, and R Heun. Confirmation of the association between bleomycin hydrolase genotype and Alzheimer's disease. *Molecular psychiatry*, 5(2):213, 2000.

Bogdan Pasaniuc, Noah Zaitlen, Huwenbo Shi, Gaurav Bhatia, Alexander Gusev, Joseph Pickrell, Joel Hirschhorn, David P Strachan, Nick Patterson, and Alkes L Price. Fast and accurate imputation of summary statistics enhances evidence of functional enrichment. *Bioinformatics*, 30(20):2906–2914, 2014.

F. Pedregosa, G. Varoquaux, A. Gramfort, V. Michel, B. Thirion, O. Grisel, M. Blondel, P. Prettenhofer, R. Weiss, V. Dubourg, J. Vanderplas, A. Passos, D. Cournapeau, M. Brucher, M. Perrot, and E. Duchesnay. Scikit-learn: Machine learning in Python. *Journal of Machine Learning Research*, 12:2825–2830, 2011.

L Peduto, VE Reuter, A Sehara-Fujisawa, DR Shaffer, HI Scher, and CP Blobel. Adam12 is highly expressed in carcinoma-associated stroma and is required for mouse prostate tumor progression. *Oncogene*, 25(39):5462, 2006.

J Perry, L Zinman, A Chambers, K Spithoff, N Lloyd, N Laperriere, Neuro oncology Disease Site Group, et al. The use of prophylactic anticonvulsants in patients with brain tumours—a systematic review. *Current Oncology*, 13(6):222, 2006.

Lukas Pezawas, Beth A Verchinski, Venkata S Mattay, Joseph H Callicott, Bhaskar S Kolachana, Richard E Straub, Michael F Egan, Andreas Meyer-Lindenberg, and Daniel R Weinberger. The brain-derived neurotrophic factor val66met polymorphism and variation in human cortical morphology. *Journal of Neuroscience*, 24(45):10099–10102, 2004.

Luke C Pilling, Jone Tamosauskaite, Garan Jones, Andrew R Wood, Lindsay Jones, Chai-Ling Kuo, George A Kuchel, Luigi Ferrucci, and David Melzer. Common conditions associated with hereditary haemochromatosis genetic variants: cohort study in uk biobank. *bmj*, 364:k5222, 2019.

G. PrasAD, T.M. Nir, A.W. Toga, and P.M. Thompson. Tractography density and network measures in Alzheimer's disease. In *Biomedical Imaging, 2013 IEEE 10th International Symposium on*, pages 692–695, 2013.

Jeffrey W Prescott, Arnaud Guidon, P Murali Doraiswamy, Kingshuk Roy Choudhury, Chunlei Liu, Jeffrey R Petrella, and Alzheimer's Disease Neuroimaging Initiative. The Alzheimer structural connectome: changes in cortical network topology with increased amyloid plaque burden. *Radiology*, 273(1):175–184, 2014.

Jesús Pujol, Anna López, Joan Deus, Narcıs Cardoner, Julio Vallejo, Antoni Capdevila, and Tomáš Paus. Anatomical variability of the anterior cingulate gyrus and basic dimensions of human personality. *Neuroimage*, 15(4):847–855, 2002.

Shaun Purcell, Benjamin Neale, Kathe Todd-Brown, Lori Thomas, Manuel AR Ferreira, David Bender, Julian Maller, Pamela Sklar, Paul IW De Bakker, Mark J Daly, et al. Plink: a tool set for whole-genome association and population-based linkage analyses. *The American Journal of Human Genetics*, 81(3):559–575, 2007.

A. Raj, E. LoCastro, A. Kuceyeski, D. Tosun, N. Relkin, M. Weiner, Alzheimer's Disease Neuroimaging Initiative (ADNI, et al. Network diffusion model of progression predicts longitudinal patterns of atrophy and metabolism in Alzheimer's disease. *Cell reports*, 10:359–369, 2015.

Eric M Reiman, Yakeel T Quiroz, Adam S Fleisher, Kewei Chen, Carlos Velez-Pardo, Marlene Jimenez-Del-Rio, Anne M Fagan, Aarti R Shah, Sergio Alvarez, Andrés Arbelaez, et al. Brain imaging and fluid biomarker analysis in young adults at genetic risk for autosomal dominant Alzheimer's disease in the presenilin 1 e280a kindred: a case-control study. *The Lancet Neurology*, 11(12):1048–1056, 2012.

KJH Robson, DJ Lehmann, VLC Wimhurst, KJ Livesey, M Combrinck, AT Merryweather-Clarke, DR Warden, and AD Smith. Synergy between the c2 allele of transferrin and the c282y allele of the haemochromatosis gene (hfe) as risk factors for developing Alzheimer's disease. *Journal of medical genetics*, 41(4):261–265, 2004.

Matthew V Rockman. Reverse engineering the genotype–phenotype map with natural genetic variation. *Nature*, 456(7223):738–744, 2008.

El Rogaev, R Sherrington, EA Rogaeva, G Levesque, M Ikeda, Y Liang, H Chi, C Lin, K Holman, T Tsuda, et al. Familial Alzheimer's disease in kindreds with missense mutations in a gene on chromosome 1 related to the Alzheimer's disease type 3 gene. *Nature*, 376(6543):775, 1995.

Roopali Roy, Ulla M Wewer, David Zurakowski, Susan E Pories, and Marsha A Moses. Adam 12 cleaves extracellular matrix proteins and correlates with cancer status and stage. *Journal of Biological Chemistry*, 279(49):51323–51330, 2004.

RStudio Team. *RStudio: Integrated Development Environment for R*. RStudio, Inc., Boston, MA, 2015. URL http://www.rstudio.com/.

M. Rubinov and O. Sporns. Complex network measures of brain connectivity: uses and interpretations. *Neuroimage*, 52(3):1059–1069, 2010.

Maurizio Salvati, Alessandro Frati, Natale Russo, Emanuela Caroli, Filippo Maria Polli, Giuseppe Minniti, and Roberto Delfini. Radiation-induced gliomas: report of 10 cases and review of the literature. *Surgical neurology*, 60(1):60–67, 2003.

Ann M Saunders, Warren J Strittmatter, D Schmechel, PH St George-Hyslop, Margaret A Pericak-Vance, SH Joo, BL Rosi, JF Gusella, DR Crapper-MacLachlan, MJ Alberts, et al. Association of apolipoprotein e allele $\epsilon 4$ with late-onset familial and sporadic Alzheimer's disease. *Neurology*, 43(8):1467–1467, 1993.

A.J. Saykin et al. Genetic studies of quantitative MCI and AD phenotypes in ADNI: Progress, opportunities, and plans. *Alzheimer's & Dementia*, 11(7):792–814, 2015.

Andrew J Saykin, Li Shen, Tatiana M Foroud, Steven G Potkin, Shanker Swaminathan, Sungeun Kim, Shannon L Risacher, Kwangsik Nho, Matthew J Huentelman, David W Craig, et al. Alzheimer's disease neuroimaging initiative biomarkers as quantitative phenotypes: Genetics core aims, progress, and plans. *Alzheimer's & dementia*, 6(3):265–273, 2010.

L Scarpace, T Mikkelsen, S Cha, S Rao, S Tekchandani, D Gutman, J Saltz, BJ ERICKSON, N PEDANO, AE FLANDERS, et al. Radiology data from the cancer genome atlas glioblastoma multiforme [tcga-gbm] collection. *The Cancer Imaging Archive*, 11:4, 2016.

Eric E Schadt, Sangsoon Woo, and Ke Hao. Bayesian method to predict individual snp genotypes from gene expression data. *Nature genetics*, 44(5):603, 2012.

Jacob Scott, Ya-Yu Tsai, Prakash Chinnaiyan, and Hsiang-Hsuan Michael Yu. Effectiveness of radiotherapy for elderly patients with glioblastoma. *International Journal of Radiation Oncology* Biology* Physics*, 81(1):206–210, 2011.

Philip Shaw, Jason P Lerch, Jens C Pruessner, Kristin N Taylor, A Blythe Rose, Deanna Greenstein, Liv Clasen, Alan Evans, Judith L Rapoport, and Jay N Giedd. Cortical morphology in children and adolescents with different apolipoprotein e gene polymorphisms: an observational study. *The Lancet Neurology*, 6(6):494–500, 2007.

Li Shen, Sungeun Kim, Shannon L Risacher, Kwangsik Nho, Shanker Swaminathan, John D West, Tatiana Foroud, Nathan Pankratz, Jason H Moore, Chantel D Sloan, et al. Whole genome association study of brain-wide imaging phenotypes for identifying quantitative trait loci in MCI and AD: A study of the ADNI cohort. *Neuroimage*, 53(3):1051–1063, 2010.

Rebecca L Siegel, Kimberly D Miller, and Ahmedin Jemal. Cancer statistics, 2019. *CA: a cancer journal for clinicians*, 69(1):7–34, 2019.

Indrapal N Singh, Robin J Goody, Celeste Dean, Nael M AhmAD, Sarah E Lutz, Pamela E Knapp, Avindra Nath, and Kurt F Hauser. Apoptotic death of striatal neurons induced by human immunodeficiency virus-1 tat and gp120: Differential involvement of caspase-3 and endonuclease g. *Journal of neurovirology*, 10(3):141–151, 2004.

Damian Smedley, Syed Haider, Steffen Durinck, Luca Pandini, Paolo Provero, James Allen, Olivier Arnaiz, Mohammad Hamza Awedh, Richard Baldock, Giulia Barbiera, et al. The biomart community portal: an innovative alternative to large, centralized data repositories. *Nucleic acids research*, 43(W1):W589–W598, 2015.

American Cancer Society. *Cancer facts & figures*. The Society, 2008.

Robert R Sokal. A statistical method for evaluating systematic relationship. *University of Kansas science bulletin*, 28:1409–1438, 1958.

Reisa A Sperling, Paul S Aisen, Laurel A Beckett, David A Bennett, Suzanne Craft, Anne M Fagan, Takeshi Iwatsubo, Clifford R Jack Jr, Jeffrey Kaye, Thomas J Montine, et al. Toward defining the preclinical stages of Alzheimer's disease: Recommendations from the national institute on aging-Alzheimer's association workgroups on diagnostic guidelines for Alzheimer's disease. *Alzheimer's & dementia*, 7(3):280–292, 2011.

Olaf Sporns. Network attributes for segregation and integration in the human brain. *Current opinion in neurobiology*, 23(2):162–171, 2013.

Olaf Sporns, Giulio Tononi, and Rolf Kötter. The human connectome: a structural description of the human brain. *PLoS computational biology*, 1(4):e42, 2005.

Tyler C Steed, Jeffrey M Treiber, Kunal Patel, Valya Ramakrishnan, Alexander Merk, Amanda R Smith, Bob S Carter, Anders M Dale, Lionel ML Chow, and Clark C Chen. Differential localization of glioblastoma subtype: implications on glioblastoma pathogenesis. *Oncotarget*, 7(18):24899, 2016.

Jason L Stein, Xue Hua, Suh Lee, April J Ho, Alex D Leow, Arthur W Toga, Andrew J Saykin, Li Shen, Tatiana Foroud, Nathan Pankratz, et al. Voxelwise genome-wide association study (vgwas). *neuroimage*, 53(3):1160–1174, 2010.

Jason L Stein, Sarah E Medland, Alejandro Arias Vasquez, Derrek P Hibar, Rudy E SenstAD, Anderson M Winkler, Roberto Toro, Katja Appel, Richard Bartecek, Ørjan Bergmann, et al. Identification of common variants associated with human hippocampal and intracranial volumes. *Nature genetics*, 44(5):552, 2012.

Gilbert Strang, Gilbert Strang, Gilbert Strang, and Gilbert Strang. *Introduction to linear algebra*, volume 3. Wellesley-Cambridge Press Wellesley, MA, 1993.

R Stupp, J-C Tonn, M Brada, G Pentheroudakis, and ESMO Guidelines Working Group. High-grade malignant glioma: Esmo clinical practice guidelines for diagnosis, treatment and follow-up. *Annals of oncology*, 21(suppl_5):v190–v193, 2010.

Damian Szklarczyk, Andrea Franceschini, Stefan Wyder, Kristoffer Forslund, Davide Heller, Jaime Huerta-Cepas, Milan Simonovic, Alexander Roth, Alberto Santos, Kalliopi P Tsafou, et al. String v10: protein–protein interaction networks, integrated over the tree of life. *Nucleic acids research*, 43(D1):D447–D452, 2014.

Damian Szklarczyk, Annika L Gable, David Lyon, Alexander Junge, Stefan Wyder, Jaime Huerta-Cepas, Milan Simonovic, Nadezhda T Doncheva, John H Morris, Peer Bork, et al. String v11: protein–protein association networks with increased coverage, supporting functional discovery in genome-wide experimental datasets. *Nucleic acids research*, 47(D1):D607–D613, 2018.

Xiaoou Tang. Texture information in run-length matrices. *IEEE transactions on image processing*, 7(11):1602–1609, 1998.

Antonio Terracciano, Toshiko Tanaka, Angelina R Sutin, Serena Sanna, Barbara Deiana, Sandra Lai, Manuela Uda, David Schlessinger, Gonçalo R Abecasis, Luigi Ferrucci, et al. Genome-wide association scan of trait depression. *Biological psychiatry*, 68(9):811–817, 2010.

Jigisha P Thakkar, Therese A Dolecek, Craig Horbinski, Quinn T Ostrom, Donita D Lightner, Jill S Barnholtz-Sloan, and John L Villano. Epidemiologic and molecular prognostic review of glioblastoma. *Cancer Epidemiology and Prevention Biomarkers*, 23(10):1985–1996, 2014.

Paul M Thompson, Nicholas G Martin, and Margaret J Wright. Imaging genomics. *Current opinion in neurology*, 23(4):368, 2010.

Paul M Thompson, Jason L Stein, Sarah E Medland, Derrek P Hibar, Alejandro Arias Vasquez, Miguel E Renteria, Roberto Toro, Neda Jahanshad, Gunter Schumann, Barbara Franke, et al. The enigma consortium: large-scale collaborative analyses of neuroimaging and genetic data. *Brain imaging and behavior*, 8(2):153–182, 2014.

Paul M Thompson, Derrek P Hibar, Jason L Stein, Gautam PrasAD, and Neda JahanshAD. Genetics of the connectome and the enigma project. In *Micro-, Meso-and Macro-Connectomics of the Brain*, pages 147–164. Springer, 2016.

Betty M Tijms, Peggy Seriès, David J Willshaw, and Stephen M Lawrie. Similarity-based extraction of individual networks from gray matter MRI scans. *Cerebral cortex*, 22(7):1530–1541, 2012.

Nathalie Tzourio-Mazoyer, Brigitte Landeau, Dimitri Papathanassiou, Fabrice Crivello, Olivier Etard, Nicolas Delcroix, Bernard Mazoyer, and Marc Joliot. Automated anatomical labeling of activations in spm using a macroscopic anatomical parcellation of the mni MRI single-subject brain. *Neuroimage*, 15(1):273–289, 2002.

Caroline Van Cauwenberghe, Christine Van Broeckhoven, and Kristel Sleegers. The genetic landscape of Alzheimer disease: clinical implications and perspectives. *Genetics in Medicine*, 18(5):421, 2016.

Marquis P Vawter, Simon Evans, Prabhakara Choudary, Hiroaki Tomita, Jim Meador-Woodruff, Margherita Molnar, Jun Li, Juan F Lopez, Rick Myers, David Cox, et al. Gender-specific gene expression in post-mortem human brain: localization to sex chromosomes. *Neuropsychopharmacology*, 29(2):373, 2004.

Roel GW Verhaak, Katherine A Hoadley, Elizabeth Purdom, Victoria Wang, Yuan Qi, Matthew D Wilkerson, C Ryan Miller, Li Ding, Todd Golub, Jill P Mesirov, et al. Integrated genomic analysis identifies clinically relevant subtypes of glioblastoma characterized by abnormalities in pdgfra, idh1, egfr, and nf1. *Cancer cell*, 17(1):98–110, 2010.

Peter M Visscher, Naomi R Wray, Qian Zhang, Pamela Sklar, Mark I McCarthy, Matthew A Brown, and Jian Yang. 10 years of gwas discovery: biology, function, and translation. *The American Journal of Human Genetics*, 101(1):5–22, 2017.

Maria Vounou, Thomas E Nichols, Giovanni Montana, Alzheimer's Disease Neuroimaging Initiative, et al. Discovering genetic associations with high-dimensional neuroimaging phenotypes: A sparse reduced-rank regression approach. *Neuroimage*, 53(3):1147–1159, 2010.

Xun-Heng Wang, Yun Jiao, and Lihua Li. Mapping individual voxel-wise morphological connectivity using wavelet transform of voxel-based morphology. *PloS one*, 13(7):e0201243, 2018.

M Watanabe, R Tanaka, and N Takeda. Magnetic resonance imaging and histopathology of cerebral gliomas. *Neuroradiology*, 34(6):463–469, 1992.

Patrick Y Wen, David R Macdonald, David A Reardon, Timothy F Cloughesy, A Gregory Sorensen, Evanthia Galanis, John DeGroot, Wolfgang Wick, Mark R Gilbert, Andrew B Lassman, et al. Updated response assessment criteria for high-grade gliomas: response assessment in neuro-oncology working group. *J Clin Oncol*, 28(11):1963–1972, 2010.

Tonya White, Jan van der Ende, and Thomas E Nichols. Beyond bonferroni revisited: concerns over inflated false positive research findings in the fields of conservation genetics, biology, and medicine. *Conservation Genetics*, pages 1–11, 2019.

Janis E Wigginton, David J Cutler, and Gonçalo R Abecasis. A note on exact tests of hardy-weinberg equilibrium. *The American Journal of Human Genetics*, 76(5):887–893, 2005.

Robert S Wilson, Patricia A Boyle, Aron S Buchman, Lei Yu, Steven E Arnold, and David A Bennett. Harm avoidance and risk of Alzheimer's disease. *Psychosomatic Medicine*, 73(8):690, 2011.

Robert S Wilson, Eisuke Segawa, Patricia A Boyle, Sophia E Anagnos, Loren P Hizel, and David A Bennett. The natural history of cognitive decline in Alzheimer's disease. *Psychology and aging*, 27(4):1008, 2012.

Anderson M Winkler, Gerard R Ridgway, Matthew A Webster, Stephen M Smith, and Thomas E Nichols. Permutation inference for the general linear model. *Neuroimage*, 92:381–397, 2014.

Kai Wu, Yasuyuki Taki, Kazunori Sato, Haochen Qi, Ryuta Kawashima, and Hiroshi Fukuda. A longitudinal study of structural brain network changes with normal aging. *Frontiers in human neuroscience*, 7:113, 2013.

Jian Yang, Beben Benyamin, Brian P McEvoy, Scott Gordon, Anjali K Henders, Dale R Nyholt, Pamela A MADden, Andrew C Heath, Nicholas G Martin, Grant W Montgomery, et al. Common SNPs explain a large proportion of the heritability for human height. *Nature genetics*, 42 (7):565, 2010.

Fiona B Young, Stefanie L Butland, Shaun S Sanders, Liza M Sutton, and Michael R Hayden. Putting proteins in their place: palmitoylation in huntington disease and other neuropsychiatric diseases. *Progress in neurobiology*, 97(2):220–238, 2012.

Richard M Young, Aria Jamshidi, Gregory Davis, and Jonathan H Sherman. Current trends in the surgical management and treatment of adult glioblastoma. *Annals of translational medicine*, 3(9), 2015.

H.-K. Ng JCS Pang MF Roussel NM Hjelm P YP Lam, E. di Tomaso. Expression of p19 ink4d, cdk4, cdk6 in glioblastoma multiforme. *British journal of neurosurgery*, 14(1):28–32, 2000.

A. Zalesky, A. Fornito, and E.T. Bullmore. Network-based statistic: identifying differences in brain networks. *Neuroimage*, 53(4):1197–1207, 2010.

Pascal O Zinn, Bhanu Majadan, Pratheesh Sathyan, Sanjay K Singh, Sadhan Majumder, Ferenc A Jolesz, and Rivka R Colen. Radiogenomic mapping of edema/cellular invasion mri-phenotypes in glioblastoma multiforme. *PloS one*, 6(10):e25451, 2011.

Chapter 7

Appendices

Table A1: Full names of brain AAL atlas regions.

Regions

	Regions	abbr.	regionLabels
Precentral_L	Precental gyrus	PreCG.L	region1
Precentral_R	Precental gyrus	PreCG.R	region2
Frontal_Sup_L	Superior frontal gyrus;dorsolateral	SFGdor.L	region3
Frontal_Sup_R	Superior frontal gyrus;dorsolateral	SFGdor.R	region4
Frontal_Sup_Orb_L	Superior frontal gyrus; orbital part	ORBsup.L	region5
Frontal_Sup_Orb_R	Superior frontal gyrus; orbital part	ORBsup.R	region6
Frontal_Mid_L	Middle frontal gyrus	MFG.L	region7
Frontal_Mid_R	Middle frontal gyrus	MFG.R	region8
Frontal_Mid_Orb_L	Middle frontal gyrus; orbital part	ORBmid.L	region9
Frontal_Mid_Orb_R	Middle frontal gyrus; orbital part	ORBmid.R	region10
Frontal_Inf_Oper_L	Inferior frontal gyrus;opercular part	IFGoperc.L	region11
Frontal_Inf_Oper_R	Inferior frontal gyrus;opercular part	IFGoperc.R	region12
Frontal_Inf_Tri_L	Inferior frontal gyrus;triangular part	IFGtriang.L	region13
Frontal_Inf_Tri_R	Inferior frontal gyrus;triangular part	IFGtriang.R	region14
Frontal_Inf_Orb_L	Inferior frontal gyrus; orbitalpart	ORBinf.L	region15
Frontal_Inf_Orb_R	Inferior frontal gyrus; orbitalpart	ORBinf.R	region16
Rolandic_Oper_L	Rolandic operculum	ROL.L	region17
Rolandic_Oper_R	Rolandic operculum	ROL.R	region18
Supp_Motor_Area_L	Supplementary motor area	SMA.L	region19
Supp_Motor_Area_R	Supplementary motor area	SMA.R	region20
Olfactory_L	Olfactory cortex	OLF.L	region21
Olfactory_R	Olfactory cortex	OLF.R	region22
Frontal_Sup_Medial_L	Superior frontal gyrus; medial	SFGmed.L	region23
Frontal_Sup_Medial_R	Superior frontal gyrus; medial	SFGmed.R	region24
Frontal_Mid_Orb_L	Superior frontal gyrus; medial orbital	ORBsupmed.L	region25
Frontal_Mid_Orb_R	Superior frontal gyrus; medial orbital	ORBsupmed.R	region26
Rectus_L	Gyrus rectus	REC.L	region27
Rectus_R	Gyrus rectus	REC.R	region28
Insula_L	Insula	INS.L	region29
Insula_R	Insula	INS.R	region30
Cingulum_Ant_L	Anterior cingulate and paracingulate gyri	ACG.L	region31
Cingulum_Ant_R	Anterior cingulate and paracingulate gyri	ACG.R	region32
Cingulum_Mid_L	Median cingulate and paracingulate gyri	DCG.L	region33
Cingulum_Mid_R	Median cingulate and paracingulate gyri	DCG.R	region34
Cingulum_Post_L	Posterior cingulate gyrus	PCG.L	region35
Cingulum_Post_R	Posterior cingulate gyrus	PCG.R	region36
Hippocampus_L	Hippocampus	HIP.L	region37
Hippocampus_R	Hippocampus	HIP.R	region38
ParaHippocampal_L	Parahippocampal gyrus	PHG.L	region39
ParaHippocampal_R	Parahippocampal gyrus	PHG.R	region40
Amygdala_L	Amygdala	AMYG.L	region41
Amygdala_R	Amygdala	AMYG.R	region42
Calcarine_L	Calcarine fissure and surrounding cortex	CAL.L	region43
Calcarine_R	Calcarine fissure and surrounding cortex	CAL.R	region44
Cuneus_L	Cuneus	CUN.L	region45
Cuneus_R	Cuneus	CUN.R	region46
Lingual_L	Lingual gyrus	LING.L	region47
Lingual_R	Lingual gyrus	LING.R	region48

Table A1: Full names of brain AAL atlas regions (continued).

Regions

Regions	Regions	abbr.	regionLab
Occipital_Mid_L	Middle occipital gyrus	MOG.L	region51
Occipital_Mid_R	Middle occipital gyrus	MOG.R	region52
Occipital_Inf_L	Inferior occipital gyrus	IOG.L	region53
Occipital_Inf_R	Inferior occipital gyrus	IOG.R	region54
Fusiform_L	Fusiform gyrus	FFG.L	region55
Fusiform_R	Fusiform gyrus	FFG.R	region56
Postcentral_L	Postcentral gyrus	PoCG.L	region57
Postcentral_R	Postcentral gyrus	PoCG.R	region58
Parietal_Sup_L	Superior parietal gyrus	SPG.L	region59
Parietal_Sup_R	Superior parietal gyrus	SPG.R	region60
Parietal_Inf_L	Inferior parietal; but supramarginal and angular gyri	IPL.L	region61
Parietal_Inf_R	Inferior parietal; but supramarginal and angular gyri	IPL.R	region62
SupraMarginal_L	Supramarginal gyrus	SMG.L	region63
SupraMarginal_R	Supramarginal gyrus	SMG.R	region64
Angular_L	Angular gyrus	ANG.L	region65
Angular_R	Angular gyrus	ANG.R	region66
Precuneus_L	Precuneus	PCUN.L	region67
Precuneus_R	Precuneus	PCUN.R	region68
Paracentral_Lobule_L	Paracentral lobule	PCL.L	region69
Paracentral_Lobule_R	Paracentral lobule	PCL.R	region70
Caudate_L	Caudate nucleus	CAU.L	region71
Caudate_R	Caudate nucleus	CAU.R	region72
Putamen_L	Lenticular nucleus; putamen	PUT.L	region73
Putamen_R	Lenticular nucleus; putamen	PUT.R	region74
Pallidum_L	Lenticular nucleus; pallidum	PAL.L	region75
Pallidum_R	Lenticular nucleus; pallidum	PAL.R	region76
Thalamus_L	Thalamus	THA.L	region77
Thalamus_R	Thalamus	THA.R	region78
Heschl_L	Heschl gyrus	HES.L	region79
Heschl_R	Heschl gyrus	HES.R	region80
Temporal_Sup_L	Superior temporal gyrus	STG.L	region81
Temporal_Sup_R	Superior temporal gyrus	STG.R	region82
Temporal_Pole_Sup_L	Temporal pole: superior temporal gyrus	TPOsup.L	region83
Temporal_Pole_Sup_R	Temporal pole: superior temporal gyrus	TPOsup.R	region84
Temporal_Mid_L	Middle temporal gyrus	MTG.L	region85
Temporal_Mid_R	Middle temporal gyrus	MTG.R	region86
Temporal_Pole_Mid_L	Temporal pole: middle temporal gyrus	TPOmid.L	region87
Temporal_Pole_Mid_R	Temporal pole: middle temporal gyrus	TPOmid.R	region88
Temporal_Inf_L	Inferior temporal gyrus	ITG.L	region89
Temporal_Inf_R	Inferior temporal gyrus	ITG.R	region90

Table A2: Top 20 Spearman association results of the change in global network metrics with targeted Alzheimer's Disease gene expressions. Threshold= $\frac{0.5}{17} = 0.0029$

Gene	Results are sorted according to p-value		
	ρ	P-value	Global Feature
PAXIP1	0.3889	0.0069	transitivity
PLAU	-0.3824	0.008	global_eff
ACE	-0.3696	0.0106	transitivity
PLAU	-0.3523	0.0151	char_path_len
ABCA7	-0.3492	0.0161	char_path_len
PSEN1	-0.299	0.0412	transitivity
APP	-0.2602	0.0774	char_path_len
PLAU	-0.2542	0.0847	louvain
APOE	-0.2506	0.0893	char_path_len
ADAM10	-0.2365	0.1095	louvain
ACE	0.2291	0.1213	louvain
NOS3	-0.2207	0.1359	char_path_len
NOS3	-0.2202	0.1369	global_eff
ABCA7	-0.2164	0.1441	global_eff
HFE	-0.2012	0.1752	char_path_len

Table A3: Top quantile regression results of the change in global network metrics and targeted Alzheimer's Disease gene expressions. Threshold= $\frac{0.5}{17} = 0.0029$.

Gene	Results are sorted according to p-value			
	Beta	Statistic	P-value	Metric
PAXIP1	0.0155	2.1179	0.039738	transitivity
PSEN1	-0.0366	-2.0821	0.04305	transitivity
A2M	0.0178	1.9728	0.054679	louvain
PLAU	-0.0157	-1.9288	0.06008	global_eff
APBB2	0.0166	1.7579	0.085573	louvain
ABCA7	0.0077	1.5185	0.135881	transitivity
BLMH	0.0101	1.3476	0.184529	transitivity
ACE	-0.0162	-1.2618	0.213524	transitivity
ADAM10	-0.0147	-1.2538	0.21638	louvain
APOE	-0.0555	-1.1673	0.249246	char_path_len
PLD3	-0.0096	-1.1433	0.258978	transitivity
ABCA7	-0.0156	-1.1367	0.261684	char_path_len
SORL1	0.0147	1.1278	0.265372	transitivity
APOE	-0.0099	-1.1048	0.275101	global_eff
HFE	-0.0161	-1.0436	0.302239	louvain

Table A4: Quantile regression top results of regressing CDR scores on the local connectivity metrics.

CDR	Results are sorted according to p-value. Threshold$=\frac{0.05}{6\times90}=9.26e-05$				
	Metric	Region	Region id	β	P-value
CDJUDGE	betweencentrality	Frontal_Inf_Oper_L	region11	-1.06e-08	1.3246e-17
CDCOMMUN	betweencentrality	Frontal_Inf_Tri_L	region13	1.162e-07	1.0377e-16
CDCOMMUN	betweencentrality	Pallidum_R	region76	6.79e-08	1.5932e-16
CDCARE	betweencentrality	Pallidum_R	region76	1.21e-08	2.5409e-15
CDCARE	betweencentrality	Frontal_Inf_Tri_L	region13	-2.35e-08	4.3817e-15
CDCARE	betweencentrality	Rolandic_Oper_R	region18	-5e-09	5.5180e-14
CDCARE	betweencentrality	Frontal_Mid_Orb_L	region9	8.35e-08	6.8455e-14
CDCARE	betweencentrality	Frontal_Inf_Tri_R	region14	1.538e-07	4.6868e-13
CDMEMORY	betweencentrality	Pallidum_R	region76	9.8e-09	1.8588e-12
CDMEMORY	betweencentrality	Heschl_R	region80	-1.431e-07	7.9135e-12
CDHOME	betweencentrality	Pallidum_R	region76	2.34e-08	1.0339e-10
CDORIENT	betweencentrality	Rolandic_Oper_R	region18	3.566e-07	2.8690e-10
CDCOMMUN	betweencentrality	Frontal_Sup_Medial_L	region23	1.2e-08	2.2249e-09
CDMEMORY	betweencentrality	Frontal_Mid_Orb_L	region9	1.53e-08	2.7792e-09
CDHOME	betweencentrality	Frontal_Inf_Tri_L	region13	7.55e-08	2.8144e-09
CDCARE	local_eff	Parietal_Sup_L	region59	-2.8362e-06	4.5150e-09
CDCARE	betweencentrality	Caudate_R	region72	-2.7e-09	5.9383e-08
CDCARE	betweencentrality	Frontal_Sup_Medial_L	region23	1.81e-08	8.4091e-08
CDCOMMUN	betweencentrality	Precentral_L	region1	5.6e-09	8.8635e-08
CDCARE	betweencentrality	Frontal_Sup_Orb_L	region5	-6.8e-09	9.1463e-08
CDCARE	betweencentrality	Occipital_Inf_L	region53	3.2e-09	1.0397e-07
CDJUDGE	betweencentrality	Insula_R	region30	-1.9e-09	1.0927e-07
CDORIENT	betweencentrality	Angular_R	region66	-2.58e-08	1.3192e-07
CDHOME	betweencentrality	Frontal_Sup_Medial_L	region23	7.28e-08	1.4609e-07
CDMEMORY	betweencentrality	Angular_L	region65	9.69e-08	1.5599e-07
CDCOMMUN	betweencentrality	Occipital_Inf_L	region53	8e-09	1.6902e-07
CDCOMMUN	betweencentrality	Occipital_Sup_R	region50	8.8e-09	1.8204e-07
CDCARE	betweencentrality	Cingulum_Post_R	region36	3.55e-08	2.6603e-07
CDCARE	betweencentrality	Angular_L	region65	5.9e-09	3.2401e-07
CDCARE	betweencentrality	Frontal_Inf_Oper_R	region12	-4.9e-09	3.9313e-07
CDMEMORY	betweencentrality	Frontal_Inf_Oper_R	region12	1.72e-08	4.2464e-07
CDJUDGE	betweencentrality	Precentral_R	region2	2.4e-09	4.2576e-07
CDJUDGE	betweencentrality	Paracentral_Lobule_R	region70	5.1e-09	4.5211e-07
CDCARE	local_eff	Caudate_L	region71	7.203e-07	6.0570e-07
CDMEMORY	betweencentrality	Occipital_Mid_R	region52	-1.45e-08	7.3061e-07
CDJUDGE	betweencentrality	SupraMarginal_R	region64	-1.35e-08	7.3933e-07
CDJUDGE	betweencentrality	Calcarine_L	region43	1.2e-09	8.0399e-07
CDCARE	betweencentrality	Precentral_L	region1	3.9e-09	1.0015e-06
CDCARE	local_eff	Thalamus_R	region78	-1.6719e-06	1.1912e-06
CDCARE	betweencentrality	ParaHippocampal_L	region39	8.1e-09	1.2821e-06
CDMEMORY	betweencentrality	Precentral_L	region1	4.9e-09	1.4549e-06
CDJUDGE	betweencentrality	Cingulum_Mid_L	region33	5e-10	1.8206e-06
CDORIENT	betweencentrality	Amygdala_R	region42	2.7e-08	2.0376e-06
CDCARE	betweencentrality	Cuneus_R	region46	2.6e-09	2.1374e-06
CDJUDGE	local_eff	Occipital_Mid_L	region51	-1.3593e-06	2.5568e-06
CDCARE	betweencentrality	Occipital_Sup_R	region50	2.4e-09	2.7811e-06
CDCARE	betweencentrality	SupraMarginal_R	region64	2.35e-08	2.8475e-06

Table A4: Quantile regression top results of regressing CDR scores on the local connectivity metrics (continued).

CDR	Results are sorted according to p-value. Threshold= $\frac{0.05}{6\times90} = 9.26e-05$				
	Metric	Region	Region id	β	P-value
CDCARE	cluster_coef	Parietal_Sup_L	region59	-1.5812e-06	3.0162e-06
CDCARE	betweencentrality	Precentral_R	region2	6.4e-09	3.0711e-06
CDJUDGE	betweencentrality	Cuneus_R	region46	8e-10	3.1560e-06
CDMEMORY	betweencentrality	Precentral_R	region2	1.21e-08	4.0628e-06
CDCARE	betweencentrality	Occipital_Mid_L	region51	-2e-09	4.7259e-06
CDCARE	betweencentrality	Temporal_Inf_R	region90	-8e-10	5.1879e-06
CDCOMMUN	betweencentrality	Temporal_Pole_Sup_R	region84	1.7e-09	5.2490e-06
CDCARE	local_eff	Paracentral_Lobule_R	region70	4.028e-07	5.3093e-06
CDCARE	betweencentrality	Olfactory_R	region22	2.5e-09	6.1963e-06
CDCARE	betweencentrality	Pallidum_L	region75	2.8e-09	6.6154e-06
CDJUDGE	betweencentrality	Postcentral_L	region57	-4e-10	6.6330e-06
CDCARE	betweencentrality	Frontal_Med_Orb_L	region25	-5.5e-09	6.7257e-06
CDCARE	betweencentrality	Parietal_Inf_L	region61	-7.6e-09	6.8700e-06
CDCARE	local_eff	Calcarine_L	region43	-1.3701e-06	6.9599e-06
CDCARE	betweencentrality	Cingulum_Mid_L	region33	9e-10	7.0795e-06
CDHOME	betweencentrality	Precentral_L	region1	6.5e-09	7.2322e-06
CDJUDGE	cluster_coef	Occipital_Mid_L	region51	-7.789e-07	7.5340e-06
CDCARE	betweencentrality	Olfactory_L	region21	1.2e-09	1.0177e-05
CDCARE	betweencentrality	Paracentral_Lobule_L	region69	2.9e-09	1.0179e-05
CDORIENT	betweencentrality	Frontal_Sup_Medial_L	region23	4.18e-08	1.1987e-05
CDCARE	cluster_coef	Thalamus_R	region78	-9.55e-07	1.2007e-05
CDMEMORY	betweencentrality	Frontal_Sup_Orb_L	region5	-2.3e-09	1.3418e-05
CDCARE	betweencentrality	Frontal_Sup_Medial_R	region24	-2.2e-09	1.3753e-05
CDMEMORY	betweencentrality	Thalamus_L	region77	4.6e-09	1.4079e-05
CDCARE	betweencentrality	Putamen_R	region74	-6.6e-09	1.4768e-05
CDCOMMUN	betweencentrality	Putamen_R	region74	-1.1e-09	1.7248e-05
CDCARE	local_eff	Amygdala_R	region42	-1.2503e-06	1.7740e-05
CDCARE	betweencentrality	Thalamus_L	region77	-7e-10	1.7820e-05
CDJUDGE	betweencentrality	Frontal_Inf_Orb_R	region16	6e-10	1.7857e-05
CDJUDGE	cluster_coef	Temporal_Mid_L	region85	6.814e-07	1.7954e-05
CDJUDGE	betweencentrality	Pallidum_L	region75	6e-10	1.9036e-05
CDCARE	local_eff	Lingual_L	region47	-1.3648e-06	1.9534e-05
CDCARE	betweencentrality	Putamen_L	region73	-9e-10	2.0543e-05
CDCARE	local_eff	Frontal_Mid_Orb_L	region9	-9.801e-07	2.0693e-05
CDJUDGE	local_eff	Temporal_Mid_L	region85	1.1769e-06	2.4200e-05
CDJUDGE	cluster_coef	Frontal_Sup_Orb_L	region5	5.54e-07	2.5902e-05
CDCARE	betweencentrality	Temporal_Pole_Sup_L	region83	-2.4e-09	2.7526e-05
CDJUDGE	local_eff	Calcarine_L	region43	-5.122e-07	2.7953e-05
CDJUDGE	cluster_coef	Cuneus_R	region46	-6.206e-07	2.8360e-05
CDCARE	betweencentrality	Frontal_Med_Orb_R	region26	1.1e-09	2.9931e-05
CDCARE	betweencentrality	Rectus_L	region27	1.7e-09	3.1086e-05
CDCARE	betweencentrality	Temporal_Pole_Sup_R	region84	4e-10	3.2473e-05
CDCARE	local_eff	Precentral_R	region2	-1.7287e-06	3.3132e-05
CDJUDGE	local_eff	Frontal_Sup_Orb_L	region5	7.069e-07	3.3295e-05
CDCARE	cluster_coef	Precuneus_L	region67	-1.3847e-06	3.3801e-05
CDJUDGE	cluster_coef	Olfactory_L	region21	3.541e-07	3.5295e-05
CDCARE	cluster_coef	Occipital_Sup_L	region49	-1.2676e-06	3.5428e-05

Table A4: Quantile regression top results of regressing CDR scores on the local connectivity metrics (continued).

CDR	Results are sorted according to p-value. $\text{Threshold} = \frac{0.05}{6 \times 90} = 9.26e - 05$				
	Metric	Region	Region id	β	P-value
CDJUDGE	cluster_coef	Calcarine_L	region43	-4.54e-07	3.7033e-05
CDJUDGE	local_eff	Cuneus_R	region46	-1.0686e-06	3.7315e-05
CDCARE	local_eff	Occipital_Sup_L	region49	-2.3923e-06	3.9588e-05
CDCOMMUN	betweencentrality	Frontal_Sup_Orb_R	region6	1e-09	4.0230e-05
CDCARE	betweencentrality	Rolandic_Oper_L	region17	6.2e-08	4.0397e-05
CDJUDGE	betweencentrality	Insula_L	region29	2e-10	4.3809e-05
CDR_diff	betweencentrality	Frontal_Sup_Medial_L	region23	0.0037021312	4.5916e-05
CDCOMMUN	cluster_coef	Precuneus_L	region67	-1.9329e-06	4.9971e-05
CDCARE	local_eff	Occipital_Mid_L	region51	-1.4926e-06	5.3963e-05
CDCARE	betweencentrality	Temporal_Inf_L	region89	2e-10	5.5267e-05
CDCARE	betweencentrality	Amygdala_L	region41	-1.1e-09	5.6693e-05
CDCARE	betweencentrality	Frontal_Inf_Orb_L	region15	-8e-10	5.6717e-05
CDJUDGE	betweencentrality	Putamen_R	region74	3e-10	5.8176e-05
CDCARE	local_eff	Parietal_Inf_L	region61	1.8772e-06	5.9063e-05
CDORIENT	local_eff	Caudate_L	region71	2.6838e-06	5.9186e-05
CDCARE	local_eff	ParaHippocampal_L	region39	-6.989e-07	6.0628e-05
CDCARE	cluster_coef	Paracentral_Lobule_L	region69	-3.682e-07	6.0769e-05
CDHOME	betweencentrality	Cingulum_Mid_L	region33	1.12e-08	6.2336e-05
CDCARE	local_eff	Temporal_Pole_Sup_R	region84	-2.5008e-06	6.4671e-05
CDCARE	betweencentrality	Calcarine_R	region44	-2.3e-09	7.4457e-05
CDCARE	betweencentrality	ParaHippocampal_R	region40	-1.2e-09	7.5724e-05
CDMEMORY	local_eff	Temporal_Mid_L	region85	4.7978e-06	7.5984e-05
CDORIENT	betweencentrality	Thalamus_L	region77	3.7e-09	8.3093e-05
CDCARE	cluster_coef	Parietal_Inf_L	region61	9.696e-07	8.3094e-05
CDCARE	betweencentrality	Precuneus_L	region67	1.3e-09	8.9269e-05
CDJUDGE	local_eff	Precentral_R	region2	-1.0795e-06	8.9400e-05
CDCOMMUN	betweencentrality	Cingulum_Mid_L	region33	1.8e-09	8.9822e-05

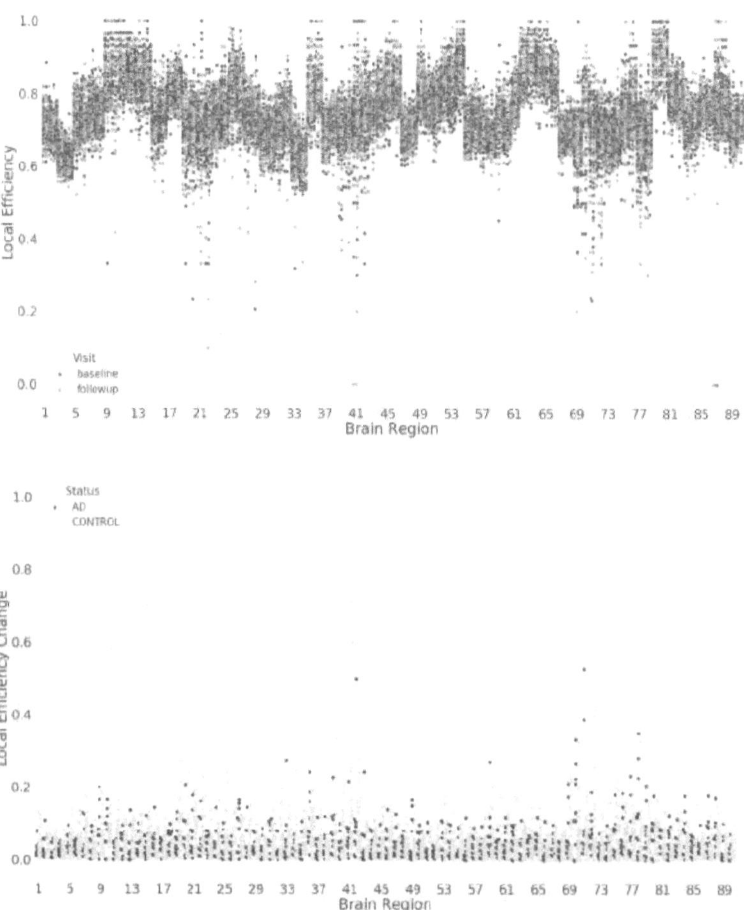

Figure A1: The top figure shows the distribution of local efficiency along the 90 AAL brain regions in the baseline (purple) vs in the follow-up (green). The bottom figure shows the distribution of the differences between the baseline and follow-up measures of local efficiency of the AD (blue) vs controls (yellow), along the 90 AAL brain regions.

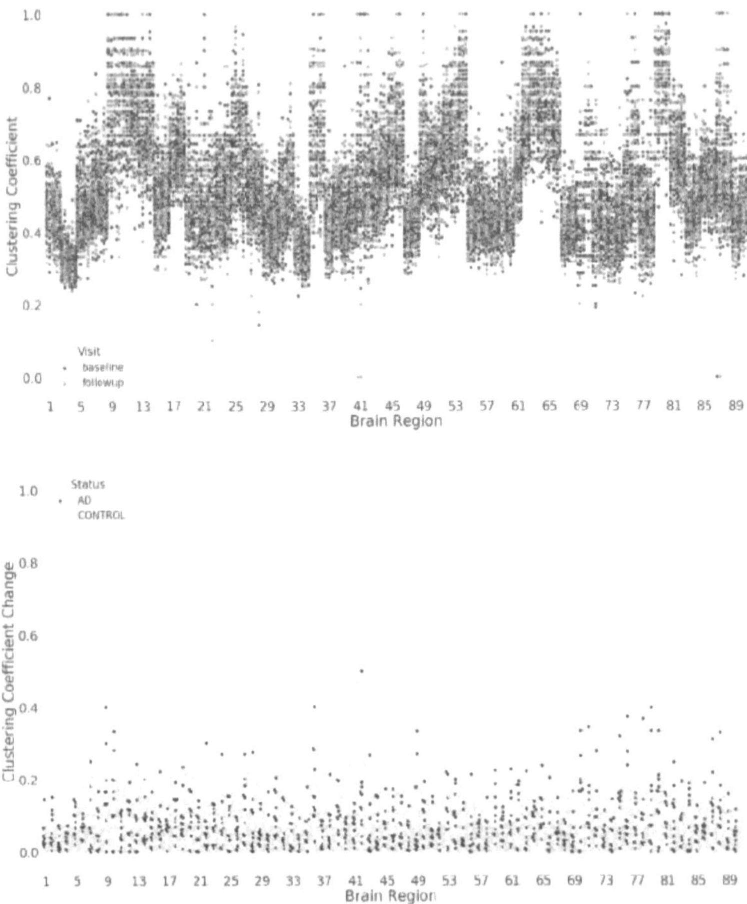

Figure A2: The top figure shows the distribution of clustering coefficient along the 90 AAL brain regions in the baseline (purple) vs in the follow-up (green). The bottom figure shows the distribution of the differences between the baseline and follow-up measures of clustering coefficient of the AD (blue) vs controls (yellow), along the 90 AAL brain regions.

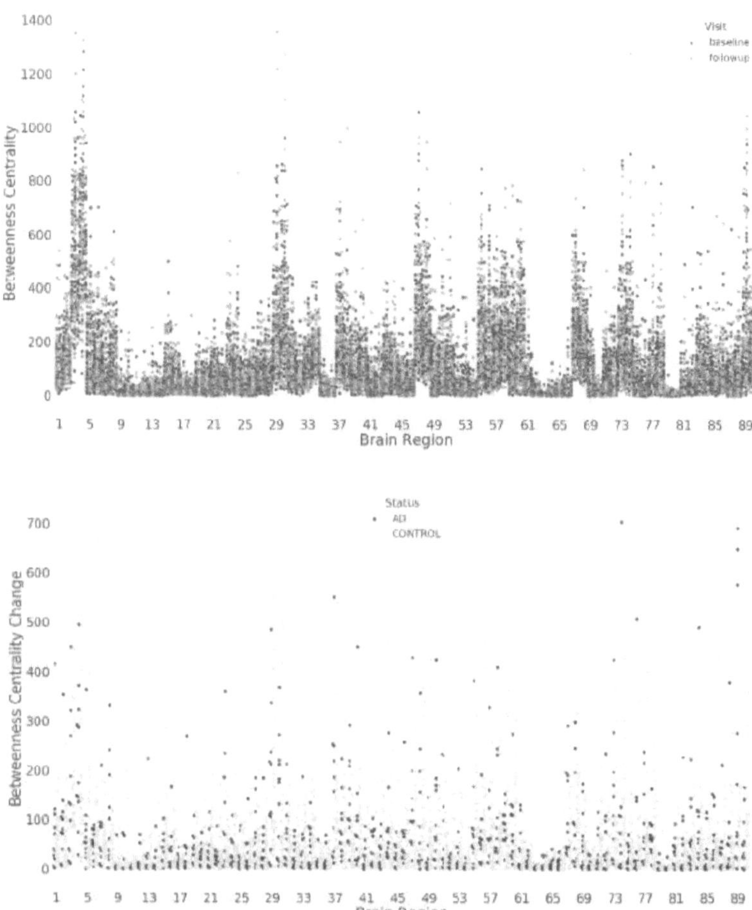

Figure A3: The top figure shows the distribution of betweenness centrality along the 90 AAL brain regions in the baseline (purple) vs in the follow-up (green). The bottom figure shows the distribution of the differences between the baseline and follow-up measures of betweenness centrality of the AD (blue) vs controls (yellow), along the 90 AAL brain regions.

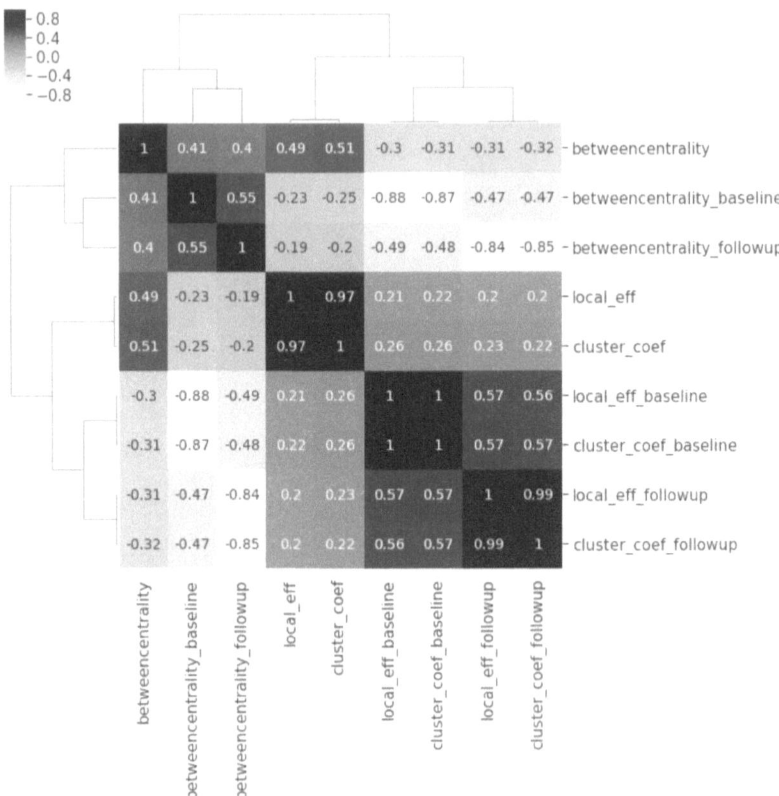

Figure A4: Spearman correlations between the three local connectivity metrics; local efficiency, clustering coefficient and betweenness centrality, at baseline (suffix: _baseline), follow-up (suffix:_followup) and the absolute difference between them (no suffix). The calculation of Sperman's coefficient combines all 90 brain regions. The plot illustrates the very strong relationship between the clustering coefficient and local efficiency at baseline, follow-up and the absolute difference between the two visits.

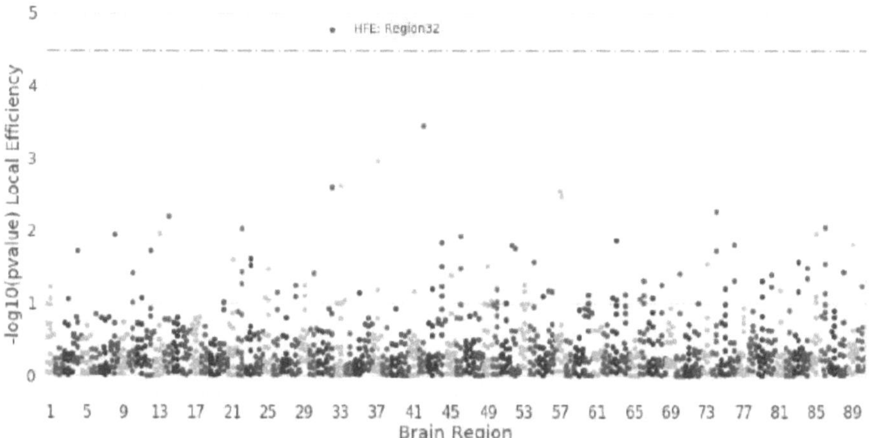

Figure A5: The figure shows the quantile regression model coefficient $-\log 10$p-values. The model regresses the change in the local coefficient (dependant variable) on a single gene at a time (independent variable), at each of the 90 brain regions as in the AAL atlas (x axis).

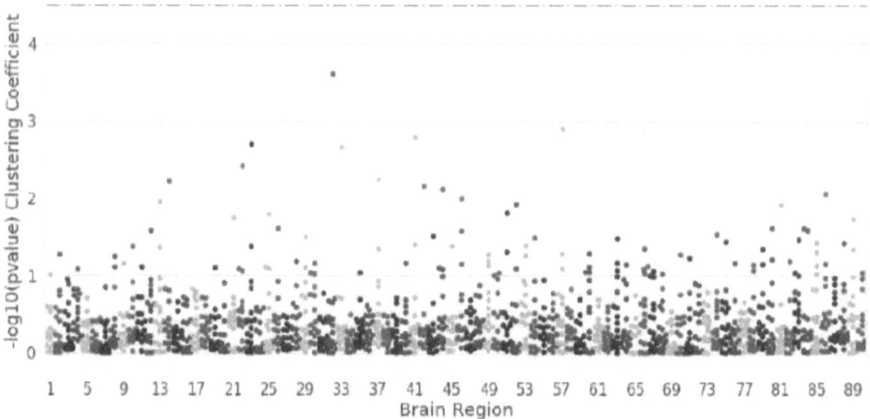

Figure A6: The figure shows the quantile regression model coefficient $-\log 10$p-values. The model regresses the change in the betweenness centrality (dependant variable) on a single gene at a time (independent variable), at each of the 90 brain regions as in the AAL atlas (x axis).

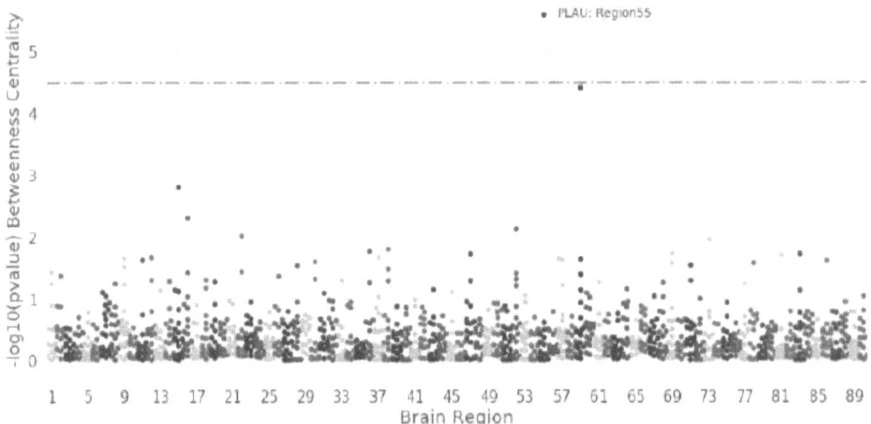

Figure A7: The figure shows the quantile regression model coefficient $-\log 10$p-values. The model regresses the change in the local coefficient (dependant variable) on a single gene at a time (independent variable), at each of the 90 brain regions as in the AAL atlas (x axis).

www.ingramcontent.com/pod-product-compliance
Lightning Source LLC
LaVergne TN
LVHW042154190726
843493LV00006B/1669